Smart Systems for Disease Prediction

Advancements, Applications and Challenges

S. Vijayalakshmi
School of Sciences
CHRIST University, India

Alwin Joseph
School of Sciences
CHRIST University, India

Naived George Eapen
School of Sciences
CHRIST University, India

Balamurugan Balusamy
Shiv Nadar University
NCR, Delhi, India

Jagjit Singh Dhatterwal
School of Computer Science & Artificial Intelligence
SR University, Warangal, Telangana, India

Kuldeep Singh Kaswan
Galgotias University
Greater Noida, UP, India

CRC Press
Taylor & Francis Group
Boca Raton London New York

CRC Press is an imprint of the
Taylor & Francis Group, an **informa** business
A SCIENCE PUBLISHERS BOOK

Cover credit: AI generated image provided by the second author, Alwin Joseph.

First edition published 2026
by CRC Press
2385 NW Executive Center Drive, Suite 320, Boca Raton FL 33431

and by CRC Press
4 Park Square, Milton Park, Abingdon, Oxon, OX14 4RN

CRC Press is an imprint of Taylor & Francis Group, LLC

Library of Congress Cataloging-in-Publication Data (applied for)

ISBN: 978-1-032-69956-1 (hbk)
ISBN: 978-1-032-72229-0 (pbk)
ISBN: 978-1-032-72234-4 (ebk)

DOI: 10.1201/9781032722344

Typeset in Times New Roman
by Prime Publishing Services

Preface

Healthcare systems have undergone huge amounts of transformations over the past decade. These changes are driven by the rapid integration of smart technologies into disease prediction, diagnosis, and management. This book offers a comprehensive overview of how Artificial Intelligence (AI), Machine Learning (ML), signal and image processing, behavioral analysis, and multimodal data processing are applied to various applications of predictive healthcare systems. The goal of this book is to bridge the gap between technological innovation and clinical applicability by providing a roadmap for researchers, healthcare professionals, and technologists in order to understand the power of smart systems for more accurate, timely, and personalized healthcare.

The chapters in this book highlight the evolving role of smart systems and how they can be used as intelligent techniques for disease prediction by integrating diverse data sources for treatment. The emphasis on intelligent methods, including deep learning, natural language processing (NLP), and multimodal data fusion, underscores the importance of combining structured and unstructured health data. The integration of multimodal data helps to analyze and provide more accurate risk estimation and preparation of personalized treatment plans. The personalized treatment plans help with addressing the complexities in modern healthcare.

The first part of the book aims to set the foundation by exploring smart systems and their role in healthcare systems. This highlights how AI-enabled tools are used to analyze vast health datasets, improving diagnostic precision and enabling clinicians to adopt data-driven decision-making practices. The chapters also provide deeper insights into specific methodologies, including the role of image and signal processing in early diagnosis, where medical imaging and physiological signals offer critical insight into the disease and its progression. The technologies and their applications help to enhance clinical workflows and allow timely detection of subtle disease markers that may go unnoticed when using traditional methods.

Another key aspect addressed is behavioral analysis, which opens new avenues for predictive healthcare by understanding lifestyle factors, social interactions, and digital footprints. This is especially relevant in the context of chronic disease prevention and mental health monitoring, where behavioral patterns serve as precursors to physiological changes. The book also discusses the impact of multimodal data processing in healthcare. This is achieved by combining electronic health records (EHRs), genomic profiles, imaging data, and real-time physiological readings. Smart systems help to enable a personalized and precision approach towards patient care.

This multimodal strategy enhances prediction accuracy by ensuring treatment plans that consider the complete health profile of each patient.

The adoption of AI in healthcare is facing challenges: the book discusses various challenges and discusses the available ethical, regulatory, and technical barriers that affect the ease of integration of smart systems in clinical practice. Challenges including data quality, interoperability, algorithmic bias, and the need for explainable AI are highlighted as critical concerns that researchers must address to build responsible AI systems. Future research direction also focuses on enhancing transparency, privacy-preserved data handling, and facilitates interdisciplinary collaboration to design systems that are both technically robust and ethically sound.

Through this overview *Smart Systems for Disease Prediction* aims to serve as a valuable resource for academicians, researchers, clinicians, and technology developers. The book offers both theoretical insights and practical frameworks to develop intelligent healthcare solutions. With the help of smart systems, healthcare can be more effective with treatment and focus on building a proactive, predictive, and personalized system that enhances the outcomes for individuals and populations alike.

Contents

1

Smart Systems in Health Care

1. Introduction to Intelligent Systems in Healthcare

Everything is changing in healthcare, as smart systems built with cutting-edge computing technology are being combined. Healthcare systems are improved with the help of Artificial Intelligence (AI) and the advent of Machine Learning (ML) and Deep Learning (DL), and data driven progonosis and healthcare systems have made the healthcare systems better and it can now foresee sickness, tailor therapies, and perhaps in time dispose of lounge waiting areas for patients altogether. Wise frameworks are showing up in medical services because of the rising intricacy of clinical information, and the requirement of individuals to settle for choices rapidly and with lower mistake rates (Jiang 2022).

Countless advancements are utilized in clever frameworks in medical services. These include computer vision, natural language processing (NLP) and prediction models. They have been developed to enable systems to look through huge files, extract useful information and help healthcare workers make smart choices. Using these tools changes how disease is diagnosed, treatments are planned and patients are managed.

One crucial area where intelligent systems have worked extraordinarily well is in predicting diseases. Predictive modeling is powered by ML algorithms. They enable us to discover health risks and diseases at a very early stage, so that we can take action before they get worse. Accordingly, 'Predictive models, for which AI is a primary ingredient, have teased out which constellation of indicators often dissolve into the beginning of long-term diseases, such as diabetes and heart disease.' AI is also being used in healthcare to manage the health of huge populations. AI has accomplished this by the collection of massive amounts of data from whole populations and analyzing that data by looking for patterns, correlations and outbreaks of viral infections so that healthcare individuals can chalk out treatment and use resources wisely so that the group's health is as optimized as possible (Wang and Jiang 2022).

DOI: 10.1201/9781032722344-1

However, it is not easy to bring clever, AI-driven systems to healthcare. There are ethical quandaries surrounding the use of medical data: for one – you don't want to compromise patient privacy or security. Balancing the benefits of these clever systems with the need to protect their data is 'essential'. To that end, researchers are exploring techniques like shared learning.

Intelligent systems are important for medical images beyond only predicting issues. Computer vision systems based on deep learning are extremely accurate when interpreting images like X-rays, MRIs and CT scans. 'If we can develop a smarter system that detects changes in the tissues and lumps that a typical doctor might otherwise oversee, then we could get the most important results to the doctor who really needs it'. This ability could significantly speed up the screening process and help to isolate problems (Ikawati et al. 2023).

EHRs have made it simple to insert smart systems into healthcare processes. NLP programs mine useful data from unstructured clinical notes. This has allowed us to learn more about the backgrounds of, and to provide more customized treatment plans.

As wearable technology and remote monitoring systems are getting better, healthcare, intelligent systems are becoming more adept. Equipped with sensors and the ability to communicate, they are bringing in a constant stream of health-related data that smart systems can keep an eye on in real-time, flagging troubling signs or sending up red flags for healthcare workers to check out.

A team of authors getting ready to publish an article about the coming age of intelligent systems in healthcare says that intelligent systems in healthcare have a promising future, but note that it will take collaboration across multiple teams to get them there. Healthcare workers, data scientists and legislators will all have to work together to keep from holding one another back, to make sure that legislation keeps up with the new tools available and to make sure these new tools can be integrated with systems that are already in place. Interdisciplinary projects are capable of overcoming problems related to data interoperability, standards, legal compliance, and more.

The idea of Explainable AI (XAI) is catching on in health care. It is crucial that intelligent systems are transparent and understandable. Making these systems explain their reasoning in ways that people can understand is crucial for building trust between health care workers and users.

1.1 Overview of Intelligent Systems

Intelligent systems are fueling a shift in healthcare from the continuous management of diseases—where a multitude of messy, unstructured health data is handled by machine learning (ML) algorithms—to one focused on diagnosing diseases and caring for patients, which involves more structured data and artificial intelligence (AI) techniques, data analytics and ML to predict the likelihood of disease, choose the best medication, or figure out the best therapy plans. A case in point is medical diagnosis, where ML techniques plow through reams of large datasets looking for patterns that might help identify diseases earlier or help radiologists read scans or pathologists review biopsies; in both cases, the images are so large that they carry huge amounts of information, making them a good target for ML techniques, which can then complement the physicians' own expertise. In image-based diagnostics, models such as convolutional neural networks (CNNs) have achieved almost unbelievable accuracy—outperforming humans in some cases—leading to better patient outcomes (Talal and Sileiman 2020).

They use predictive analytics—AI allows these intelligent systems to monitor a patient's health around the clock, looking at what is happening with them in context with what has happened in the past, learning to notice small details that a human might not recognize, and guessing what might happen next. That enables doctors to take action before a patient gets too sick for them to help the patient. Electronic health records (EHRs) make unstructured clinical accounts full of useful data available at a doctor's fingertips. Simple inquiries like these have now become significantly more complex due to advancements in data processing and intelligent systems. Natural language processing (NLP) systems built into EHRs can quickly pull up the needed information from unorganized clinical accounts and help healthcare workers make better decisions (Choudhary et al. 2021).

Smart tools are now being used in healthcare for more than just remote patient tracking. Wearable monitors collect real-time bodily data, which lets doctors keep an eye on someone and act quickly if needed. Astute calculations are utilized by Clinical Choice Emotionally Supportive Networks (CDSS) to assist medical care suppliers with settling on decisions in view of realities. The mix of patient information and clinical data in these frameworks make conclusion and treatment designs more precise (Huang et al. 2019).

Automated medical procedure utilizes shrewd frameworks since it should be exceptionally exact and skillful. Consolidating robots with man-made intelligence improves procedures by expanding accuracy and diminishing recuperation periods. Despite their enormous potential, intelligent systems are difficult to use in healthcare due to ethical considerations, the need for standards, and data security. The utilization of these advances relies upon the goal of these issues.

For astute frameworks to work appropriately, it is vital to guarantee that different medical service information sources can speak with each other and team up without issues. Endeavors to lay out principles and guarantee interoperability have extraordinary importance in this specific situation. We should stress what the utilization of artificial intelligence advances in medical services means for moral contemplations. Concerns like understanding security, informed assent, and algorithmic predisposition should be entirely examined to lay out validity and the assurance of proper use.

The advancement of intelligent frameworks has worked extraordinarily with customized medication. Customized treatment methodologies that record for individual varieties are made conceivable by these advancements, which analyze DNA, clinical data, and way-of-life information. Advancing healthcare relies heavily on genomic data, and smart systems are capable of efficiently processing and analyzing vast amounts of genomic information. Structured genetic data analysis enhances diagnostic accuracy and facilitates the identification of personalized treatment options (Shalini 2021).

Smart frameworks are truly adept at having the option to change. These frameworks stay aware of the most recent clinical examinations and innovative advances by utilizing processes that let them learn and develop over the long haul. As healthcare systems become more interconnected, protecting private medical data is more critical than ever. Strong cybersecurity measures are required to safeguard patient data and maintain the proper operation of healthcare AI systems.

By taking a gander at organic information to find conceivable medication possibilities, savvy frameworks assume a major part in drug improvement. This accelerates the drug development process, allowing new treatments to reach the market more quickly. Telemedicine stages, which are controlled by shrewd frameworks, make it more straightforward to get clinical consideration and engage in talk from a distance. These systems make healthcare more accessible, especially in areas that are not easily accessible, and open up new ways to customize care (Srilakshmi et al. 2019).

It is very important for clever systems and healthcare workers to work together. Working together with AI improves clinical knowledge, with AI giving useful insights and helping to make decisions. Clear rules are needed because clever systems are being used more and more in healthcare. Regulatory bodies are changing to make sure that AI technologies used in medicine are safe, effective, and moral. Patients are given more power by intelligent tools that give them access to personalized health information (Alahmadi and Soh 2011). Engaging patients with health tech encourages them to take an active role in their care and helps them make decisions together with their caregivers. As clever systems keep getting better, the future looks bright for more progress in healthcare. For these systems to work at their best and improve patient results, researchers, healthcare workers, and technologists must work together.

1.2 Importance in Healthcare

The use of smart systems, which use advanced technologies to improve disease forecasts, analysis, and patient care, is changing healthcare in a big way. As these systems are very important for data analysis because they give real-time information about how patients are doing and how their treatments are working. Smart systems use ML methods to look at huge amounts of data and predict diseases. This lets them find possible health problems early on. Predictive analytics helps with preventive healthcare methods, which makes things easier for both patients and healthcare workers (Abdeen et al. 2021).

The Internet of Things (IoT) has changed the way patients are monitored by letting data be collected and sent all the time. Wearable devices and sensors track vital signs, giving healthcare workers fast information for effective action. Personalized medicine depends on smart systems that make treatment plans based on collected data about each patient. A big step forward in patient-centered care has been made possible by this accuracy in healthcare, which reduces side effects and improves therapeutic results.

Smart systems have made telehealth and online patient tracking more popular, especially since global events have made it harder to provide healthcare in person. These tools make it easier for people to get medical care, especially in rural or underserved areas. Blockchain technology is being used more and more to protect health data, making sure that patient safety and data security are maintained (Safari et al. 2018). Trust in healthcare systems grows because the blockchain is open and can't be changed. This makes electronic health records safer.

Adding smart systems to medical images has made it easier and more accurate to find diseases. Computer programs that use AI look at medical pictures and help doctors find small problems that people might miss. With the help of smart systems, health informatics makes it easy for healthcare workers to share patient information, which encourages them to work together and organize care. This connectivity makes healthcare service more efficient as a whole.

Using robotic systems in surgery, guided by AI, has improved accuracy and decreased the amount of damage done. Robotic-assisted treatments are becoming more common because they have better results and shorter healing times. During pandemics, predictive models and data analytics have been very important in public health. They help predict how diseases will spread and where resources should be used. Smart systems help with statistical studies, which help policymakers make decisions based on facts (Xanthidis et al. 2020).

Smart systems help keep an eye on and predict mental illnesses by looking at trends in behavior and bodily data. These tools help with early prevention and personalized mental health care. Smart systems make it possible for electronic health records to talk to each other, which lets doctors see full patient histories and make better decisions. The availability of an all-around view of patient data makes it easier for care to stay consistent across different healthcare settings.

Healthcare robots that use natural language processing and machine learning keep patients more interested and give them information quickly. These talking bots help people make appointments, remember to take their medications, and learn about health. Smart systems add to the development of pharmacogenomics, tailoring drug doses based on genetic information and this personalized method reduces bad drug reactions and improves therapeutic results.

Smart systems look at huge amounts of data to find trends. This speeds up the process of finding new treatments and solutions. These tools make clinical studies more efficient and help move medicine toward evidence-based practice. AI-driven decision support systems help doctors make good choices by looking at patient data and suggesting the best ways to treat patients. These systems are useful in complicated medical situations because they help doctors make more accurate diagnoses and plan better treatments.

Smart systems in healthcare add to the early discovery of infectious diseases through real-time tracking of epidemiological data. This method lets public health officials step in at the right time to stop the spread of dangerous diseases. Smart systems in healthcare have ethical effects, such as privacy issues and algorithmic bias, that require ongoing review and the creation of strong regulatory frameworks.

Making sure that these technologies are used in a fair and reasonable way is important for making sure that they continue to improve patient results (Kumar et al. 2023).

2. Applications of Intelligent Systems in Healthcare

Intelligent systems are very important for finding diseases because they use complex algorithms and machine learning to look at huge amounts of data. Pattern recognition features in these systems help with early evaluation, which leads to better solutions. Using smart systems to find diseases means combining different types of data, like electronic health records and personal tech. These systems allow real-time analysis because they are constantly watching, which helps find health problems quickly (Ismanov et al. 2018). In healthcare, ML systems help figure out how bad a disease is by looking at a number of clinical factors. To figure out how bad diseases are and how fast they're spreading, these systems look at things like biomarker levels, imaging data, and the patient's medical background. Intelligent systems can assess disease severity in personalized medicine, enabling the creation of treatment plans tailored to each patient's unique characteristics. This method improves the accuracy of healthcare treatments, which leads to better results.

Prognostic systems that use AI make a big difference in healthcare by figuring out how diseases will progress. These systems use predictive models to guess how diseases will get worse, which lets doctors plan the right treatments. Intelligent systems in healthcare not only find diseases, but also help doctors figure out how bad infectious illnesses are, which is very important for managing public health well. Analyzing the seriousness of a disease is especially important for long-term conditions, where smart systems can help with ongoing tracking and changing treatment plans to lower risks.

Machine learning models utilize patient data, lifestyle habits, and genetic information to predict the severity and progression of cardiovascular diseases, enabling early intervention and personalized treatment planning. Cancer prognostic systems use genetic data to make personalized estimates about how the disease will progress and how well the treatment will work; intelligent systems can help neurologists figure out how bad neurological diseases are by analyzing data from neuroimaging and patient-reported results. Using smart systems to find diseases is making progress in infectious disease situations, making it easier to quickly identify germs and target treatments. Intelligent systems are also utilized in mental health to assess severity by analyzing patterns in speech, behavior, and other data. These algorithms enable early detection of mental health issues and facilitate timely interventions.

Prognostic systems are an important part of managing diabetes because they help doctors figure out how to treat patients so they have better results. By automatically analyzing medical pictures, intelligent systems in radiology make it easier to find diseases. This cuts down on diagnosis time and improves accuracy. Smart systems that track physiological factors help with disease intensity analysis in respiratory illnesses. This allows for proactive management and lowers the number of hospitalizations. Predictive methods for infectious diseases help with public health planning by predicting how diseases will spread and finding places where they might be most likely to happen.

Predicting how bad rare diseases will be requires advanced analytics because there isn't a lot of data available. Intelligent systems are very important for this. Smart disease detection systems are enhanced by integrating personal health technologies, which provide continuous health data for early warning and timely intervention. Severity analysis in autoimmune diseases benefits from intelligence systems studying immunological markers, helping in personalized treatment strategies. As multi-modal data is added to healthcare prognostic systems, genetic, clinical, and environmental factors are being combined to make more accurate disease predictions (Chiuchisan et al. 2014).

3. Focus on Predictive Systems in Healthcare

With the rise of prediction systems in the last few years, the healthcare industry has undergone a major change. To fully understand prediction systems in healthcare, you need to know how to use complex computer models to look at very large data sets. These systems use ML algorithms, statistical models, and AI to guess what will happen, what the trends will be, and what health risks might be there for patients (Guo et al. 2016).

Predictive modelling has a lot of potential in healthcare because it lets doctors plan ahead for how to care for patients. These systems can predict possible health problems before they happen by looking at past data, present health situations, and different factors that might affect them. Healthcare professionals are able to intervene early, thereby improving long-term patient outcomes, by adopting this proactive strategy. Predictive systems can optimize resource utilization in healthcare organizations by enabling proactive planning for periods of increased patient demand.

Prescient displaying assumes a pivotal part in medical care since it upgrades the accuracy of conclusion. By incorporating a few information sources like electronic wellbeing records, hereditary information, and way of life information, these frameworks might give an exhaustive outline of a patient's wellbeing. This exhaustive methodology empowers clinical experts to settle on additional educated conclusions about analysis and treatment procedures.

Expectation frameworks have the worthwhile ability of upgrading the particularity of medication. Medical care experts might upgrade treatment techniques by creating individualized therapy plans as indicated by the interesting qualities and chance variables of every patient. The accentuation on quiet focused approach improves the adequacy of treatment, yet in addition lessens pointless aftereffects and techniques (Ranjith and Mahantesh 2019).

Prescient displaying assumes an urgent part in the protection of medical services. These strategies engage doctors and medical caretakers to recognize people who have a high likelihood of fostering specific sicknesses. The medical services framework might alleviate the transmission of diseases and improve the general prosperity of the entire local area through designated intercessions, for example, altering people's propensities or working with early screenings.

It is becoming apparent that the utilization of gauging devices may significantly aid the administration of persistent diseases as their viability moves along. These frameworks have the capacity to consistently screen a patient's wellbeing and give continuous information on the movement of the disease and the viability of the treatment. Continuous monitoring allows for timely adjustments to treatment strategies, enhancing the adaptability and effectiveness of chronic disease management.

Prescient demonstrating is valuable for both general wellbeing drives and individual patient treatment. These systems may be able to identify patterns and trends that can assist in the early detection of clusters or illnesses by analyzing comprehensive population data. Fast reaction is vital in restricting the spread of contaminations, especially in circumstances when the capacity to answer quickly is fundamental (Martinez-Balleste 2019).

Prescient advances are additionally being utilized in a few medical care settings outside clinical settings. These models might help medical services managers advance asset designation, smooth out techniques, and improve the by and large functional effectiveness of their organizations. This far-reaching approach improves the medical services climate, consequently advancing a more powerful and feasible framework.

Predictive systems offer several advantages; however, they also present challenges that require resolution for optimal implementation and effectiveness. Defending the classification and security of patient information is of utmost importance, requiring the execution of strong measures to guarantee the insurance of private data. Ensuring ethical considerations remain a priority throughout the development and implementation of these systems is crucial, requiring a careful balance between innovation and responsible use.

The development of forecast frameworks in medical services is intensely attached to progressions in innovation. The capacities of these frameworks might improve as handling power and calculation intricacy keep on progressing. The consolidation of arising advancements, for example, edge figuring and the Internet of Things (IoT) upgrades the possibility of medical services mediations being information driven and happening quickly.

When putting predictive systems into use in healthcare, it is essential for experts from different fields to work together. Working together with experts from a few fields, for example, wellbeing, software engineering, information investigation, and morals guarantees the technique's breadth and all-encompassing methodology. Cooperative endeavors empower the foundation of normalized standards and ideal procedures. This improves the viability and cohesiveness of prescient displaying in medical care (Eeshwaroju et al. 2020).

For expectation frameworks to accomplish far-reaching reception, they should have the capacity to be handily changed regarding scale. It is essential to ensure that prediction models are capable of effectively processing the vast amount of data due to the growing size and complexity of healthcare records. Healthcare organizations, from small clinics to large hospital networks, require scalable solutions that can be easily adapted to meet their specific needs.

In healthcare, prediction modeling plays a crucial role in assisting clinicians in making decisions. These frameworks upgrade the dynamic course of medical services experts by giving them proof-based data and suggestions, empowering them to pursue more educated choices sooner rather than later. This expansion of human navigation might bring about superior results for patients and a more smoothed out way to deal with medical services conveyance.

To gather certainty and regard inside the medical care business, forecast models should have an elevated degree of fathomability. Considering that these models frequently work obscurely, upgrading their straightforwardness and comprehensibility is pivotal. The understanding of the medical services staff of the components behind prescient frameworks improves their trust and utilization of such frameworks in their normal assignments.

Table 1. Predictive systems in healthcare.

Aspect	**Description**
Application Areas	1. Disease Prediction and Prevention 2. Patient Risk Stratification 3. Treatment Personalization 4. Resource Optimization 5. Early Diagnosis and Intervention
Technologies Used	1. Machine Learning (ML) 2. Artificial Intelligence (AI) 3. Predictive Analytics 4. Data Mining 5. Natural Language Processing (NLP)
Data Sources	1. Electronic Health Records (EHR) 2. Genomic Data 3. Wearables and IoT Devices 4. Imaging and Diagnostic Tests 5. Social Determinants of Health (SDOH)
Challenges	1. Data Privacy and Security Concerns 2. Integration of Diverse Data Sources 3. Interoperability Issues 4. Ethical Considerations 5. Limited Explainability of AI Models
Success Metrics	1. Improved Patient Outcomes 2. Reduction in Hospital Readmissions 3. Increased Accuracy of Diagnoses 4. Enhanced Healthcare Resource Utilization 5. Cost Savings and Efficiency Gains
Examples of Systems	1. IBM Watson for Oncology 2. Google's DeepMind Health 3. Predictive Analytics for Heart Disease 4. Diabetic Retinopathy Prediction Models 5. Population Health Management Platforms
Regulatory Considerations	1. FDA Approval for AI-Based Medical Devices 2. Compliance with HIPAA Regulations 3. Ethical Guidelines for AI in Healthcare 4. Transparent Reporting and Accountability 5. Continuous Monitoring and Auditing of Predictive Models

Although predictive systems hold a lot of promise, they must constantly be tested and improved to ensure their accuracy and dependability. These models continue to get better over the long run thanks to steady audit and changes in view of how they work in reality. Using prediction models in healthcare is supported by a solid scientific foundation, which is consistent with evidence-based medicine (Table 1).

As healthcare evolves, predictive systems will play a crucial role in shaping the future. The way patients are cared for, how resources are used, and how well healthcare generally works could all be completely altered as these new technologies are introduced gradually. To capitalize on expectation frameworks in medical services, it is vital to go into this change with care, tackling issues, ensuring that morals are considered, and empowering individuals to cooperate (Bhatt and Chakraborty 2021).

4. Intelligent Systems for Disease Prediction

Huge steps have been taken in the right direction in medical services with regard to utilizing AI frameworks to anticipate illnesses. AI techniques are pivotal since they empower the advancement of savvy frameworks that possibly upset sickness forecasts. The use of various machine learning models, such as support vector machines, neural networks, and ensemble approaches to improve the accuracy and efficacy of illness forecasting has been the subject of extensive research.

Using prescient displaying approaches is essential in utilizing canny frameworks for sickness expectation. These methodologies utilize authentic patient information, clinical records, and other applicable information to gauge the likelihood of disease events. Using prescient demonstrating works with early issue recognition, additionally upgrades modified medication by fitting medical care therapies to the particular prerequisites of individual patients.

Information driven techniques are fundamental in astute medical services frameworks for ailment expectation. Electronic wellbeing records, clinical pictures, and hereditary data incorporate significant measures of information, which have incited the advancement of many-sided information driven models. AI calculations succeed in distinguishing examples and patterns related with different diseases when given broad and various datasets, bringing about additional exact expectations.

The utilization of clever advancements in medical services fundamentally changes the course of disease expectation. These frameworks utilize a few AI procedures to oversee enormous datasets and extricate important data for prescient expectation. Researchers have examined the utilization of calculations, for example, choice trees, irregular timberlands, and profound learning in the distinguishing proof of perplexing examples in medical care information. This has worked due to the improvement of strong models for sickness expectation (Zenkert et al. 2022).

By employing novel methods to identify disorders, machine learning algorithms surpass conventional statistical methods in healthcare. Bayesian organizations give a measurable structure to addressing mind boggling collaborations inside medical care information. This approach upgrades the understandability of sickness expectation models, furnishing clinicians with significant experiences with the factors that impact assessments.

The essential target of building complex medical care frameworks is to empower ongoing sickness forecast. By coordinating ongoing information with web-based learning approaches, these frameworks have the capacity to adjust quickly as per advancing medical services conditions. Not only does the use of real-time illness prediction make it easier for people to get help quickly, but it also makes predictive models more accurate and effective over time.

The meaning of considering morals in the utilization of shrewd frameworks for disease expectation can't be overemphasized. As these advances gain expanding importance in the medical care area, it is basic to resolve issues connected with protection, security, and informed assent. To guarantee capable use of wise frameworks in medical services, it is necessary to strike a sensitive balance between the benefits of exact sickness forecast and the need to defend the privileges of patients.

Regardless of headways in involving modern frameworks for disease expectation, there are annoying issues. Incorporating divergent data from a few sources could hinder their cooperative working. To improve the utilization of wise frameworks in medical services, it is critical to take a stab at the normalization of information organizations and interoperability strategies.

Reasonable man-made intelligence (XAI) is acquiring prevalence as a pivotal part of canny medical services frameworks. For sickness expectation models to acquire the confidence of medical care staff and the overall population, they should have an elevated degree of understandability. Reasonable Man-made brainpower (XAI) devices, for example, model-rationalist techniques and rule-based clarifications, upgrade the straightforwardness of AI models. This works with perception and upgrades the soundness of the results inside a helpful setting. Adding subject data to shrewd frameworks that anticipate sicknesses makes them work better. AI calculations and master data can be joined in half breed models to make forecasts that are more precise and consider what is happening. Consolidating subject information with information driven strategies creates a strong blend that can deal with even the most convoluted medical care circumstances.

For astute frameworks to be generally utilized in medical services, they should have the option to be extended. As medical care information continues to develop at a remarkable rate, it is essential to have frameworks that can deal with records of all sizes and levels of intricacy. A few innovations that can make it more straightforward for sharp frameworks for infection conjecture to deal with a bigger scope are distributed computing, circulated registering, and equal handling.

It is vital to zero in on individuals while making brilliant medical care frameworks. The user interfaces of decision-support tools should be simple and easy to comprehend for healthcare professionals. Involving end users in the design process ensures that intelligent systems are not only accurate but also practical and seamlessly integrated into existing healthcare practices.

For insightful frameworks to find lasting success at anticipating illnesses, they need to cooperate with individuals from various fields, like officials, medical services laborers, and PC researchers. It is essential to link technical knowledge with specific field knowledge in order to develop solutions that are both scientifically sound and applicable to the real world. Problems in healthcare forecast can be examined more thoroughly when people collaborate.

To guarantee unwavering quality, shrewd frameworks utilized for ailment expectation should reliably confirm and approve their models. Given the powerful idea of medical care information, it is fundamental to update and approve models to guarantee their precision and appropriateness in assorted situations. Utilizing strong approval strategies is fundamental for upgrading and refining prescient models in the unique medical care climate.

Laying out a thorough structure for information management is the basic requirement for using clever frameworks for illness identification. A powerful information administration structure is important to protect security, stick to moral standards, and guarantee consistence with lawful necessities. Sticking to laid out conventions guarantees that the utilization of wise frameworks in medical care isn't just successful but also additionally agreeable with lawful and moral standards.

Highlighting choice methodologies assumes a significant part in getting exact figures about sicknesses. Recognizing significant qualities among the immense amount of information is vital for creating proficient and successful models. Feature selection approaches, such as filtering, wrapping, and embedding techniques, enhance the performance of intelligent systems by prioritizing the most informative components.

The job of large information examination in anticipating sicknesses in shrewd frameworks is profoundly praiseworthy. Because of the immense and complicated nature of medical services information, it requires the utilization of modern investigation instruments fit for distinguishing stowed away examples and associations. Joining enormous information examinations with AI calculations makes it feasible for shrewd frameworks to get valuable data from huge and differed datasets.

With regards to anticipating infections, gathering techniques functions admirably. In healthcare settings, ensemble learning, which improves the ability of multiple models to predict the future by combining them has gained popularity. By making guesses that can be used in a variety of healthcare situations, bagging and boosting methods increase the reliability of intelligent systems.

A complete way to deal with hacking is required on the grounds that brilliant frameworks are being utilized to foresee illnesses. As healthcare data is increasingly stored digitally, ensuring the security and privacy of sensitive information is more critical than ever. To keep patient information protected and of shrewd frameworks in medical services dependable (Figure 1), it is essential to utilize encryption, secure specialized strategies, and use ordinary security checks.

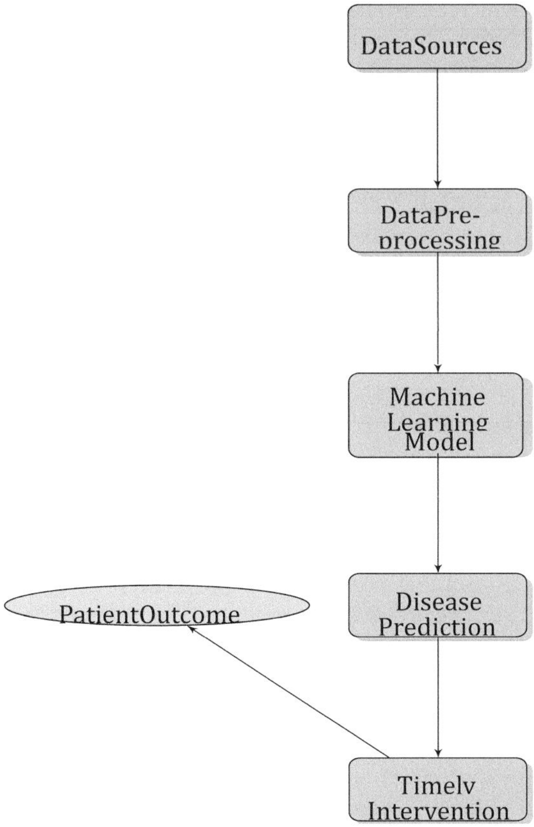

Figure 1. Intelligent systems for disease prediction.

5. Case Studies and Examples

Utilizing astute frameworks to foresee sicknesses in reality has shown that they can have an impact on how medical care is finished. One intriguing contextual analysis sees how AI strategies can be utilized to early track down of diabetic retinopathy.

Profound learning models were utilized to take a gander at photos of the retina and sort out how likely it was that diabetic retinopathy would frame. Effectively instituting these savvy frameworks prompted more precise early finding of the condition, which took into account speedy treatment and better quiet outcomes.

An intriguing case study demonstrates how predictive modeling techniques can be utilized to identify individuals who are likely to develop heart diseases in the context of cardiovascular health. The savvy framework had the option to partition individuals into bunches in view of the fact that they were very liable to have a cardiovascular occasion. It did this by utilizing a great deal of electronic wellbeing records and various kinds of information, like hereditary data and data about an individual’s way of life. As indicated, this strategy makes centered preventive advances simpler, which helps bring down the cases of cardiovascular grimness and demise.

One significant model in medication is the utilization of information driven strategies to anticipate bosom disease. Smart frameworks with cutting edge AI calculations took a gander at mammograms and patient foundation data to sort out how likely it was that somebody would get bosom disease. At the point when these frameworks were instituted and functioned admirably, early spotting rates went up immensely. This implied that treatment could begin all the more rapidly, which further reduced patient death rates.

A contextual investigation on the utilization of gathering techniques in irresistible illness expectation shows how well they work at foreseeing sickness episodes. To take a gander at epidemiological measurements and outer elements, specialists joined different AI models, for example, choice trees and backing vector machines. The savvy framework accurately anticipated how irresistible infections would spread, which helped general wellbeing authorities make explicit moves to stop breakouts before they occurred.

Utilizing shrewd frameworks to sort out who is probably going to get Alzheimer's sickness is another strong model. Analysts utilized progressed AI strategies to make forecast models by assembling neuroimaging information, hereditary data, and mental tests. These models were truly adept at finding individuals who were probably going to get Alzheimer's sickness. This made it feasible for early intercession strategies and explicit treatment plans.

Intelligent systems have demonstrated promise for improving diagnosis accuracy for rare diseases. Hereditary sequencing information and clinical data were utilized together for a situation learn about an interesting hereditary infection. The AI calculations effectively recognized inconspicuous examples demonstrative of the unprecedented sickness. This aided doctors in making exact analyses and fitting treatments for people distressed by the condition.

A dataset that contains a well-organized data that is publically available in Kaggle datasets relating to cardiovascular issues. The documents incorporate a few standards, including age, circulatory strain, cholesterol levels, maximal pulse, and other relevant realities. Specialists and information researchers utilize these datasets to dissect patterns, relationships, and foster expectation models in the field of cardiovascular wellbeing. Different datasets, working with their availability and use for anyone with any interest at all in doing research and studies relating to cardiovascular wellbeing.

In medical care datasets, Table 2 regularly lists the numerous classifications of information used to give clinical data. These factors incorporate unmitigated factors like patient socioeconomics (e.g., orientation and identity), mathematical factors like crucial signs (e.g., pulse and pulse), transient variables (e.g., confirmation and release dates), and text-based factors (e.g., clinical notes and findings). For healthcare information organization and analysis, such data is essential. Furthermore, they work with the administration of information, empower informed direction, and upgrade research abilities inside the perplexing domain of medical care frameworks.

Table 2. Datatypes in healthcare datasets.

Data columns (total 14 columns):			
#	**Column**	**Non-Null Count**	**Dtype**
0	age	1025 non-null	int64
1	sex	1025 non-null	int64
2	cp	1025 non-null	int64
3	trestbps	1025 non-null	int64
4	chol	1025 non-null	int64
5	fbs	1025 non-null	int64
6	restecg	1025 non-null	int64
7	thalach	1025 non-null	int64
8	exang	1025 non-null	int64
9	oldpeak	1025 non-null	float64
10	slope	1025 non-null	int64
11	ca	1025 non-null	int64
12	thal	1025 non-null	int64
13	target	1025 non-null	float64

dtypes: float64(2), int64(12)
memory usage: 112.2 KB

Table 3 contains different medical services datasets including list, age, sex, cp (sort of chest torment), trestbps (resting pulse), chol (serum cholesterol), fbs (fasting glucose), restecg (resting electrocardiographic outcomes), thalach (most extreme pulse accomplished), exang (work out instigated angina), oldpeak (ST discouragement initiated by practice comparative with rest), slant, ca (number of significant vessels shaded by fluoroscopy), thal (thalassemia), and target (pr). Each line in the information contains important data for medical services exploration and examination, addressing an unmistakable patient and their comparing values for these traits.

Table 3. Healthcare datasets.

Index	Age	Sex	Cp	Trestbps	Chol	Fbs	Restecg	Thalach	Exang	Oldpeak	Slope	Ca	Thal	Target
0	52	1	0	125	212	0	1	168	0	1.0	2	2	3	0.23
1	53	1	0	140	203	1	0	155	1	3.1	0	0	3	0.37
2	70	1	0	145	174	0	1	125	1	2.6	0	0	3	0.24
3	61	1	0	148	203	0	1	161	0	0.0	2	1	3	0.28
4	62	0	0	138	294	1	1	106	0	1.9	1	3	2	0.21

Figure 2 shows a dissipate plot that shows the connection among age and the objective variable. It shows how the information is fanned out and whether there are any potential examples. Every information point is a recording of a solitary individual, and the age values are plotted against the objective qualities that go with them. The scatter plot shows any clear patterns or links between age and the goal

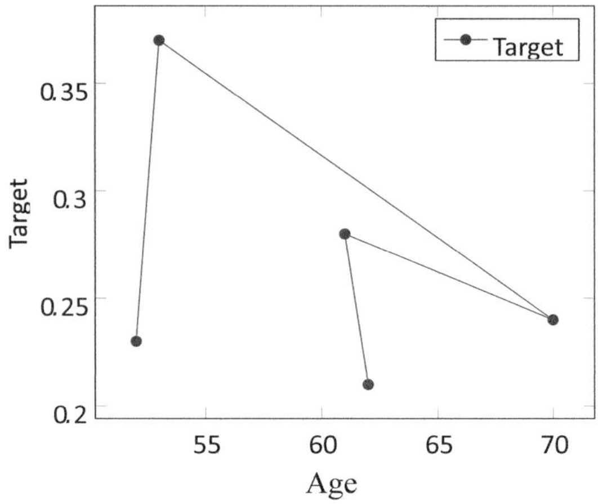

Figure 2. Scatter plot of age vs target.

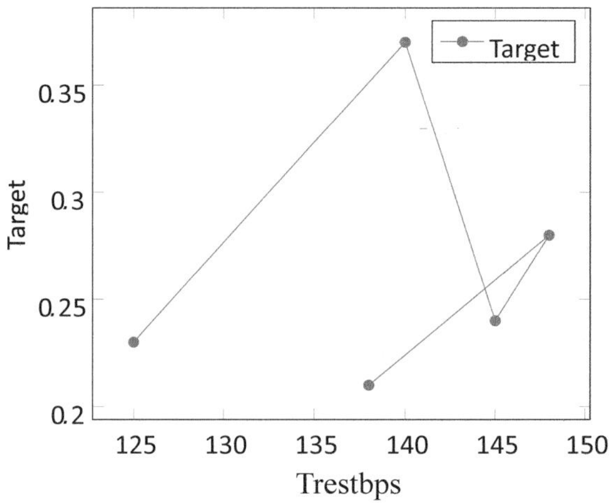

Figure 3. Scatter plot of trestbps vs target.

variable. It can be used as a visual help for further analysis and interpretation within the research study or dataset.

A scatter plot, which shows the connection between trestbps (resting blood pressure) and the goal variable, is shown in Figure 3. The pattern of data points spread out across the plot's axis shows the relationship. The main purpose of the visualization is probably to look for any possible link or trend between resting blood pressure and the desired result. It does this by giving useful information about the features of the dataset. The way in which the points are spread out can show groups, outliers, or the lack of a clear pattern. This may provide crucial clues for future model evaluation and research in this particular project or study.

6. Challenges and Future Directions

While brilliant frameworks in medical services have a ton of potential, they likewise have an issue that should be fixed before they can be utilized in a viable and moral manner. A general issue is the absence of interoperability across medical services frameworks, bringing about different information designs and a failure to speak with each other. The coordination of different information sources might be trying because of their absence of interoperability. This is a test in creating coordinated and thorough prescient medical care frameworks.

While creating and involving wise frameworks in medical care, focusing on morals and information security is essential. The utilization of private clinical information for disease expectation raises moral worries around consent, straightforwardness, and the suitable administration of individual data. To build trust between doctors and patients, it is important to strike a balance between protecting people's privacy and the potential benefits of planned treatment.

The inherent complexity of comprehending machine learning models is one major issue with anticipated healthcare systems. Every now and again, complicated calculations, for example, profound learning models work as 'secret elements', making it hard to perceive the hidden components behind their interesting expectations. Guaranteeing the fathomability of models is essential for accumulating acknowledgement from medical care experts and working with informed dynamics in clinical settings.

The potential predispositions in medical care information, which smart frameworks might further enhance and spread, are a separate problem. Changing socioeconomics may achieve dissimilar levels of exactness in making forecasts, while the preparation information shows a predisposition. Eliminating these biases is essential for preventing health disparities from worsening and ensuring equitable healthcare outcomes.

Patterns and enhancements in prescient medical care frameworks for the future point towards various fascinating paths. Logical simulated intelligence (XAI) is probably going to turn out to be more significant as endeavors continue to make AI models more obvious. New XAI strategies, similar to demonstrating rationalist interpretability and consideration instruments, are intended to assist us with better seeing how savvy frameworks simplify decisions.

Predictive healthcare systems should work better in real-time if they can connect to new technologies like edge computing and 5G networks. Edge computing moves the handling of data closer to the source, which cuts down on delay and speeds up decision-making. When these technologies are combined with clever systems, they could change how patients are monitored remotely and decide when help is needed.

It is believed that collaborative and cross-disciplinary study will grow, making it easier for data scientists, healthcare professionals, and policymakers to work together. By working together, the difference between technical know-how and domain-specific knowledge can be filled. This ensures that predictive healthcare systems are not only technically sound but also fit the needs of healthcare workers and the rules that govern them.

New developments in federated learning and autonomous models may help with privacy issues by letting machine learning models be trained to spread data without affecting the privacy of individual data. This method lets students from different schools work together to learn while having private data in one place.

7. Conclusion

When smart systems are used to predict diseases, they change everything about healthcare and start a new age of accurate and proactive medical treatments. Advanced machine learning methods, predictive modeling techniques, and data-driven approaches have made disease forecasts much more accurate and efficient for a wide range of medical conditions. Intelligent systems have shown they can look at huge amounts of data, find small trends, and make very accurate predictions about the chance of diseases, from heart diseases to rare genetic disorders. Some important things we learned about intelligent systems for disease prediction are that they can make a huge difference in how well patients do by finding problems early, making personalized treatment plans, and taking specific preventative measures.

The given case studies show successful applications in various medical areas, showing real improvements in the accuracy of diagnoses and the prompt start of interventions. These real-life examples show how important clever systems are for changing how healthcare is provided and making everyone healthier. For intelligent systems to reach their full potential, problems like data sharing, social issues, and model interpretability must be solved by teams of people from different fields working together. Future developments in predictive healthcare systems, such as improvements in AI that can be explained, integration with new technologies, and joint research efforts, could help get around the current problems and push the limits of what is possible in disease prediction.

Intelligent systems are more than just new technology. They are also useful tools that help healthcare workers make smart choices, give better care, and save lives. As the field of predictive medicine changes, it will be important to keep looking for ethical, open, and patient-centered ways to use intelligent systems in disease prediction in a responsible and effective way. We are still moving toward a better future, thanks to the positive path that these smart systems have set out.

References

Abdeen, M., Ahmed, M. H., Seliem, H., El-Nainay, M. and Sheltami, T. R. (2021). Improving the performance of ambulance emergency service using smart health systems. 2021 IEEE/ACM Conference on Connected Health: Applications, Systems and Engineering Technologies (CHASE), Washington, DC, USA, pp. 205–209, doi: 10.1109/CHASE52844.2021.00045.

Alahmadi, A. and Soh, B. (2011). A smart approach towards a mobile e-health monitoring system architecture. 2011 International Conference on Research and Innovation in Information Systems, Kuala Lumpur, Malaysia, pp. 1–5, doi: 10.1109/ICRIIS.2011.6125669.

Bhatt, V. and Chakraborty, S. (2021). Real-time healthcare monitoring using smart systems: A step towards healthcare service orchestration Smart systems for futuristic healthcare. 2021 International Conference on Artificial Intelligence and Smart Systems (ICAIS), Coimbatore, India, pp. 772–777, doi: 10.1109/ICAIS50930.2021.9396029.

Chaudhary, R., Kumar, R., Chatterjee, K. and Singh, A. (2021). A health monitoring framework based on edge of things for smart healthcare system. 2021 IEEE 6th International Conference on Computing, Communication and Automation (ICCCA), Arad, Romania, pp. 339–344, doi: 10.1109/ICCCA52192.2021.9666367.

Chiuchisan, I., Costin, H. -N. and Geman, O. (2014). Adopting the Internet of Things technologies in health care systems. 2014 International Conference and Exposition on Electrical and Power Engineering (EPE), Iasi, Romania, pp. 532–535, doi: 10.1109/ICEPE.2014.6969965.

Eeshwaroju, S., Jakkula, P. and Ganesan, S. (2020). IoT based empowerment by smart health monitoring, smart education and smart jobs. 2020 International Conference on Computing and Information Technology (ICCIT-1441), Tabuk, Saudi Arabia, pp. 1–5, doi: 10.1109/ICCIT-144147971.2020.9213754.

Guo, Y., Yang, L., Zhu, H., Cheng, Y., Tian, F. and Zhao, S. (2016). Smart health service system: objectives, framework and solution. 2016 International Computer Symposium (ICS), Chiayi, Taiwan, pp. 680–684, doi: 10.1109/ICS.2016.0140.

Huang, P. -C., Lin, C. -C., Wang, Y. -H. and Hsieh, H. -J. (2019). Development of health care system based on wearable devices. 2019 Prognostics and System Health Management Conference (PHM-Paris), Paris, France, pp. 249–252, doi: 10.1109/PHM-Paris.2019.00049.

Ikawati, V., Saepudin, D., Susanto, B., Munandar, A., Nurba, H. P. and Prihatmanto, A. S. (2023). Individual health profiling using multimodal sensors on smart health care. 2023 17th International Conference on Telecommunication Systems, Services, and Applications (TSSA), Lombok, Indonesia, pp. 1–5, doi: 10.1109/TSSA59948.2023.10366913.

Ismanov, Z., Iversen, M., Kaasbøll, J., Manda, T. and Ni, I. (2018). Using patient information systems for primary health care in hospitals – implications for data structures and user interaction. 2018 IST-Africa Week Conference (IST-Africa), Gaborone, Botswana, pp. Page 1 of 10–Page 10 of 10.

Jiang, Y. (2022). Construction of an intelligent system for elderly's health and elderly care from the perspective of the integration of smart sensors and physical medicine. 2022 4th International Conference on Smart Systems and Inventive Technology (ICSSIT), Tirunelveli, India, pp. 541–544, doi: 10.1109/ICSSIT53264.2022.9716434.

Kumar, K., Chaudhary, S., Anandaram, H., Kumar, R., Gupta, A. and Joshi, K. (2023). Industry 4.0 and health care system with special reference to mental health. 2023 1st International Conference on Intelligent Computing and Research Trends (ICRT), Roorkee, India, pp. 1–5, doi: 10.1109/ICRT57042.2023.10146640.

Martínez-Ballesté, A., Gimeno, P., Mariné, A., Batista, E. and Solanas, A. (2019). e-PEMICU: an e-Health platform to support early mobilisation in intensive care units. 2019 10th International Conference on Information, Intelligence, Systems and Applications (IISA), Patras, Greece, pp. 1–6, doi: 10.1109/IISA.2019.8900718.

Ranjith, J. and Mahantesh, K. (2019). Privacy and security issues in smart health care. 2019 4th International Conference on Electrical, Electronics, Communication, Computer Technologies and Optimization Techniques (ICEECCOT), Mysuru, India, pp. 378–383, doi: 10.1109/ICEECCOT46775.2019.9114681.

Safari, G., Majidi, B., Khanzadi, P. and Manzuri, M. T. (2018). Cross-platform E-management for smart care facilities using deep interpretation of patient surveillance videos. 2018 25th National and 3rd International Iranian Conference on Biomedical Engineering (ICBME), Qom, Iran, pp. 1–6, doi: 10.1109/ICBME.2018.8703545.

Shalini, V. B. (2021). Smart health care monitoring system based on Internet of Things (IOT). 2021 International Conference on Artificial Intelligence and Smart Systems (ICAIS), Coimbatore, India, pp. 1449–1453, doi: 10.1109/ICAIS50930.2021.9396019.

Srilakshmi, A., Mohanapriya, P., Harini, D. and Geetha, K. (2019). IoT based smart health care system to prevent security attacks in SDN. 2019 Fifth International Conference on Electrical Energy Systems (ICEES), Chennai, India, pp. 1–7, doi: 10.1109/ICEES.2019.8719236.

Talal, R. and Suleiman, A. R. (2020). IoT/IoE smart health care system using fog computing features. 2020 2nd Annual International Conference on Information and Sciences (AiCIS), Fallujah, Iraq, pp. 28–34, doi: 10.1109/AiCIS51645.2020.00015.

Wang and Jiang, L. (2022). Research on the information construction plan of smart health and senior care villages. 2022 8th Annual International Conference on Network and Information Systems for Computers (ICNISC), Hangzhou, China, pp. 904–906, doi: 10.1109/ICNISC57059.2022.00182.

Xanthidis, D., Manolas, C., Xanthidou, O. K., Adnan, O. and Jamal, K. (2020). Smart health: sensors, big data, and cloud computing in the health sector in the U.A.E. 2020 24th International Conference on Circuits, Systems, Communications and Computers (CSCC), Chania, Greece, pp. 45–51, doi: 10.1109/CSCC49995.2020.00017.

Zenkert, J., Karl, D., Dornhöfer, M., Fathi, M. and Becker, C. (2022). LOKAL-digital - smart knowledge management for housing, care and health. 2022 IEEE International Smart Cities Conference (ISC2), Pafos, Cyprus, pp. 1–4, doi: 10.1109/ISC255366.2022.9921847.

2

Intelligent Techniques for Disease Prediction

1. Introduction to Intelligent Techniques in Disease Prediction

At a global level, healthcare systems are being revolutionized, and adopting smart methods for predicting illnesses and mitigating their effect has become common. A powerhouse of smart systems for disease prediction has been created by combining advanced computer science with health care. Artificial intelligence and machine learning in these systems have successfully predicted many diseases early (Babu and Karthick 2022).

These intelligent methods, which are based on extremely large datasets, discover intricate patterns and relationships for millions of patients, which would be impossible using traditional diagnostic approaches. The smart system behind patient data integrates genetic information and environmental factors, utilizing complex formulas to produce highly accurate and insightful medical analyses. This provides a complete description of the beginning and spreading process of diseases. Smarter suggestions emerge that help detect all illnesses, from chronic diseases to infectious diseases, and emerging health threats. These systems are agile and can be deployed across various healthcare settings. This may well translate to a paradigm shift of preventive healthcare tactics as we know it now (Lakshmanarao et al. 2021).

Despite much promise, there are problems with such clever methods finding their way into healthcare. Ensuring the privacy and security of private health data is crucial. In addition, complicated formulas are difficult for healthcare workers to understand and reveal moral issues. The heart of the issue is that current strategies for predicting disease do not work well. The traditional ways of working with complex and ever-evolving contemporary healthcare data usually don't cut. That's where intelligent techniques with flexibility and learning come to be a hopeful solution for overcoming such limitations.

The primary motivation behind the addition of smart methods to disease prediction models is largely accuracy and utility. With the ability to detect diseases earlier, these tools can be used for prompt interventions and cause-specific individualized treatments that often lead to improved patient outcomes. The overarching premise

DOI: 10.1201/9781032722344-2

is the transition of healthcare management from passive to active. This part refers to advancements, applications and limitations of artificial intelligence in disease prediction. We seek to demonstrate the value of smart solutions in changing healthcare delivery by examining numerous case studies and the latest research.

The interesting thing about smart hacks is that they can evolve with new lessons of life. The more these systems process data and the longer they are allowed to learn, the better they become at making accurate predictions and informed guesses. Healthcare adapts very quickly, so disease prediction models always improve. Smart techniques are not solely used in diagnostics, but also for evaluating the risks and developing preventive measures. Smart systems facilitate healthcare professionals to target their interventions on groups and individuals with high risk in order to address the absolute disease burden (Sowmiya and Sumitra 2017).

Remembering that smart systems in healthcare are inherently complex and span multiple fields is the key. The most powerful way to create and apply ingenious techniques of disease identification is the collaboration between computer scientists, healthcare professionals or some other types of experts in this domain with data processing. When surveying clever methods or interventions, you have to be as skeptical of the successes of any method as its failures. Finally, we aim to show what these methods look like in the real world and how well they perform via empirical analysis that help us understand disease forecast. The application of intelligent methods for predicting diseases becomes a turning point in the history of health. The purpose of this chapter is to provide you with an overview on all new and recent development in the field and their uses, highlighting some major challenges faced by showcasing relevant recent works from Lafta et al. 2015.

Health care is undergoing a massive change with clever methods being added continuously. Thanks to artificial intelligence (AI) and machine learning (ML), these systems have revolutionized how doctors diagnose patients and predict patient outcomes. The advancement of technology is essential since millions of heavy medical data can be easily analyzed by advanced systems and provide precise information to predict disease better.

The last few years have seen a big shift in predicting diseases with the advances of clever systems. Complex algorithms analyze a multitude of datasets—from genetic information, to medical records and lifestyle factors. The ability to bring in so many different types of data, and link them directly or indirectly (roughly 200 people turn up with chicken pox every day), helps automatic systems uncover intricate patterns and connections for better disease predictions quicker.

Within healthcare, one of the things that smart systems can offer above all else is early indicators and preventative measures from contracting different diseases. From COVID-19 exposure tracking to symptoms prediction, the systems can predict early on and allow healthcare workers some time to manage patients effectively which may lead to less severe illnesses and thus better outcomes. Noteworthy, early identification may change the course of long-term conditions and planning for this is particularly significant in chronic disorders (KS et al. 2020).

Second, real-time clever systems are going to be way more useful for predicting disease. Since the health of a patient's system is always monitoring them and analyzing data, these systems can adapt to changes in their health situation. That way, forecasts stay accurate and current. Real-time tools like these can be invaluable, particularly in the case of infectious diseases and other emerging health risks.

Another critical function that highly adept intelligent systems can perform is the personalization of healthcare treatments. Customized predictions and recommendations for each patient are obtained with these systems since the models hunt down variations in their genetics, living styles as well as environmental elements. This individualized strategy not only makes disease predictions more precise, but it also improves the effectiveness of treatment plans.

However, there are issues when it comes to deploying smart tools over research and healthcare. Maintaining the security and confidentiality of patient data is a continued concern. Ensuring trust in these smart systems requires maintaining data integrity while prioritizing preserving and securing patients' healthcare information.

AI systems have recently been designed to help doctors better interpret AI-driven predictions, enhancing clinical decision-making and diagnostic accuracy. The need continues to grow as intelligent systems are integrated to generate their suggestions, particularly when their decision-making processes are critical. However, ensuring transparency and reliability is essential for building trust and acceptance within the medical community. Confidence in these systems is crucial for their widespread adoption and effectiveness in healthcare. Closing this interpretability gap is critical to seamlessly integrate intelligent systems into existing healthcare workflows (Raja et al. 2016).

Regardless, it cannot be denied that intelligence systems are critical for healthcare. Each device excels in utilizing AI to predict diseases, proving valuable not only for individual patient care but also for broader public health applications. That means a huge potential for proactive, data driven personalized medicine in there. Such a future would be within grasp by incorporating intelligent tools into healthcare workflows.

2. Overview of Disease Prediction Methods

Brilliant frameworks, which are fueled by PC knowledge, have become key apparatuses in changing how illnesses are anticipated. These frameworks utilize many techniques, from antiquated measurable models to state-of-the-art AI calculations. They are vital for further developing medical services. Utilizing these savvy methods together is important for early location and treatment of sicknesses (Franklin and Muthukumar 2020).

Predictive modeling, which employs statistical techniques to examine previous data and identify trends that can inform us about how health will be in the future, is an essential component of disease prediction (Singh and Kaswan 2024). AI calculations, particularly profound learning models, have become famous in light of the fact that they can naturally gain convoluted designs from exceptionally huge datasets. This improves the accuracy of disease forecasts. In this time of 'huge information', enormous medical services datasets have made it more straightforward for information driven techniques to recognize sicknesses. These strategies utilize various information sources, for example, electronic wellbeing records, hereditary data, and information from brilliant gadgets, to give a full image of every individual's wellbeing profile (Sharma and Singhal 2023).

With the ascent of accuracy medication, techniques for anticipating infections have moved much further into the universe of customized medical services. By joining hereditary and living information about a solitary patient, brilliant frameworks make it conceivable to make customized forecast models that consider how vulnerable every individual is to disease. In sickness expectation, shrewd frameworks accomplish something beyond check out the strength of an individual. They likewise check out the wellbeing of an entire local area. PC based epidemiological displaying can anticipate sickness bunches, which makes it more straightforward for general healthcare authorities to act rapidly (Priyanga and Naveen 2017).

Predicting diseases is now much simpler thanks to sophisticated tools that combine a variety of medical scans. The utilization of AI to see clinical pictures like X-ray and CT filters works on the exactness of determinations and helps find illnesses early. A vital piece of shrewd medical services frameworks is utilizing continuous information streams to further develop models that foresee illnesses. Wearable gadgets that continually track physiological variables permit the structure of dynamic forecast models that give precise data about changes in wellbeing (Amin et al. 2013).

Cooperating across various fields is vital for making and working on savvy frameworks for anticipating sicknesses. Utilizing information from software engineering, medication, and information science together ensures that model creation and assessment are finished in a total manner. Despite the fact that there have been enhancements, it is still difficult to utilize savvy frameworks to anticipate sicknesses, in actual medical care circumstances. Information security, the simplicity of deciphering confounded models, and moral worries all should be painstakingly explored.

For illness estimate strategies to be solid and valuable in different circumstances, they should be assessed and approved constantly. Savvy frameworks work better in an assortment of medical services settings when they are thoroughly tried against various datasets and genuine circumstances. Prediction models need to be simple to be comprehended by people and for healthcare workers to trust them. Settling on model choice making processes ensures that shrewd frameworks are utilized in medical care circumstances in a moral way.

Using explainable artificial intelligence (XAI) techniques, the issue of comprehending intricate smart systems is resolved. XAI techniques assist us with sorting out how models make specific expectations, which expands trust and acknowledgment in medical services settings. Shrewd frameworks should have the option to be extended to be generally utilized in an assortment of medical care settings. Ensuring that prescient models can work with various sums and sorts of information makes it more straightforward to involve them in medical care processes (Rane 2018).

It is vital to fix mistakes in medical care information since savvy frameworks are turning into a significant piece of foreseeing sicknesses. It is critical to do a full survey and concoct ways of fixing one-sided datasets, so out of line results don't occur. Since brilliant frameworks change so rapidly, they should be refreshed and changed constantly to keep them aware of new medical services issues. Standard model refreshing and improvement is expected to ensure that illness forecast techniques continue to function admirably.

For shrewd frameworks to be utilized in medical care, administrative guidelines need to change. To ensure that prescient models are utilized in a moral manner that safeguard patients' inclinations, clear standards and rules are required. For shrewd frameworks to be utilized mindfully in illness figures, they need to cooperate with government bodies, organizations, and colleges. These kinds of agreements can make it simpler to establish standardized procedures and frameworks for the validation and implementation of models (Geeitha et al. 2023).

2.1 Traditional Disease Prediction Approaches vs. Intelligent Techniques

As we study Table 1 we see that there is an evident difference between two methods of prediction for diseases. The table compares the traditional and intelligent methods. This is where smart methods such as machine learning, Artificial Intelligence (AI), and big data analytics come into play to have a better forecast at a far quicker rate than traditional ways of purely looking towards statistical models or easy formulas. The smart approaches can easily handle the complex and non-stationary health-care data when compared with traditional approaches. It will change how illnesses are predicted and make more customized healthcare solutions possible.

2.2 Evolution of Intelligent Techniques in Disease Prediction

Table 2 studies changes in smart techniques for predicting diseases over the years. It demonstrates how economical statistical models to complex AI driven solutions are being developed. Disease forecasting has been made more accurate and efficient over the years thanks to machine learning, artificial intelligence, big data analytics, etc. This chapter presents a history of these technologies and how they evolved. The table outlines how these intelligent methods have revolutionized healthcare, notably by facilitating the development of individual treatment plans with higher quality and expediting more accurate disease outbreak prediction as well as requiring patient-care data.

Table 1. Comparing traditional disease prediction approaches and intelligent techniques.

Aspect	Traditional Disease Prediction Approaches	Intelligent Techniques
Data Source	Limited data sources, primarily manual input	Diverse data sources, including electronic health records, genetic data, wearable devices, etc.
Data Preprocessing	Manual and rule-based preprocessing	Automated and adaptive preprocessing using machine learning algorithms
Feature Selection	Basic features manually selected	Advanced feature selection using algorithms like Recursive Feature Elimination, feature importance from machine learning models
Model Complexity	Simple models (e.g., linear regression)	Complex models (e.g., ensemble methods, deep learning)
Expert Knowledge	Highly reliant on domain experts	Can learn patterns and features without explicit expert input
Adaptability	Limited adaptability to changing scenarios	Adaptive and can learn from new data patterns, adjusting to changing conditions
Interpretability	Models may lack interpretability	Models can provide insights into decision-making processes, improving interpretability
Handling non-linearity	Limited capability to handle non-linear relationships	Able to capture complex non-linear relationships through advanced modeling techniques
Scalability	Limited scalability with large datasets	Scalable to handle large datasets, parallel processing, and distributed computing
Real-time Processing	Often lacks real-time prediction capabilities	Capable of real-time processing, critical for timely decision-making
Prediction Accuracy	May have lower accuracy due to simplicity	Can achieve higher accuracy by learning intricate patterns in the data
Automation	Limited automation of the prediction process	Automation of the entire prediction pipeline, from data preprocessing to model deployment
Integration of Multiple Modalities	Limited integration of various data modalities	Can integrate diverse data types (e.g., images, text, numerical data) for comprehensive analysis
Transfer Learning	Rarely implemented	Utilizes transfer learning to leverage knowledge gained from one domain for improved predictions in another domain

Table 2. Evolution of intelligent techniques in disease prediction.

Time Period	Intelligent Techniques	Key Features
Pre-2000s	Traditional Statistics	Basic statistical models for disease risk assessment.
2000s	Rule-Based Systems	Expert systems using predefined rules and knowledge bases.
	Machine Learning	Initial use of algorithms like Decision Trees and Support Vector Machines.
	Genetic Algorithms	Incorporation of evolutionary algorithms for optimization.
2010s	Deep Learning	Introduction of neural networks for feature learning. Convolutional Neural Networks (CNNs) for image-based diagnostics.
	Ensemble Methods	Combination of multiple models for improved accuracy.
	Transfer Learning	Leveraging pre-trained models on related tasks for improved performance.
	Feature Engineering	Improved feature extraction for better model input.
2020s	Explainable AI (XAI)	Emphasis on interpretability for clinical acceptance.
	Reinforcement Learning	Dynamic adaptation of models based on feedback loops.
	Integrative Approaches	Integration of multi-modal data for comprehensive analysis.
	Personalized Medicine	Tailoring predictions based on individual patient characteristics.
Future	Quantum Computing	Potential for handling vast datasets and complex computations.
	Advanced AI Ethics	Greater focus on responsible and ethical use of AI in healthcare.

3. Predictive Modeling in Disease Prediction

Adding smart systems to healthcare is changing everything, especially when it comes to using prediction models to stop diseases before they happen. Machine learning systems have changed everything by giving us new ways to look at patient data and how diseases change over time.

3.1 Application of Machine Learning Algorithms

When it comes to predicting diseases, machine learning systems are by far the most useful tools. Support Vector Machines (SVMs) and Random Forests are great at making predictions because they can find complex trends in a wide range of datasets. Deep learning methods, such as neural networks, have also shown amazing skills in dealing with complicated health data, which has led to better prediction accuracy (Ghazal et al. 2022).

There are a lot of different medical situations where machine learning can be used to identify diseases. In particular, cancer prediction algorithms like Logistic Regression and Decision Trees have done a better job of predicting how patients will do based on genetic and clinical data. Similar to this, infectious disease cases have been successfully identified using models that take into account environmental factors and spread rates (Vij and Kaswan 2023).

Looking at expectation models for various infections is conceivable with brilliant frameworks, which assists us with completely figuring out their assets and shortcomings. For instance, in cardiovascular circumstances, correlation studies have seen how well various strategies anticipate risk factors, giving data about which model to use for which gatherings of patients (Djerioui et al. 2020).

3.2 Comparative Analysis of Predictive Models in Various Diseases

Correlation incorporates neurological issues, where AI techniques have been utilized to distinguish how the sickness will deteriorate and assist with making customized therapy arrangements. Predictive models based on brain data have shown promise in the early detection and monitoring of Alzheimer's disease, which is one important application.

For instance, models like SEIR (Powerless Uncovered Irresistible Eliminated) are utilized to think about how infections like Coronavirus will spread and demonstrate the way that adaptable prescient displaying can be. The relative trial of these models demonstrates the way that they can foresee the future, which assists administrators to work out ways of controlling infection (Choudhary et al. 2023).

The comparison of different disease prediction models is shown in Figure 1. It shows how well they work for a number of different illnesses. It shows how well different models work, such as basic statistical models, machine learning methods, and advanced artificial intelligence techniques. The graph shows that AI methods are better at predicting and treating most diseases, which shows how they could completely change how diseases are diagnosed and treated. This also shows that disease prediction is changing all the time, from using old methods to using more modern, data-driven models that give us more information and better estimates.

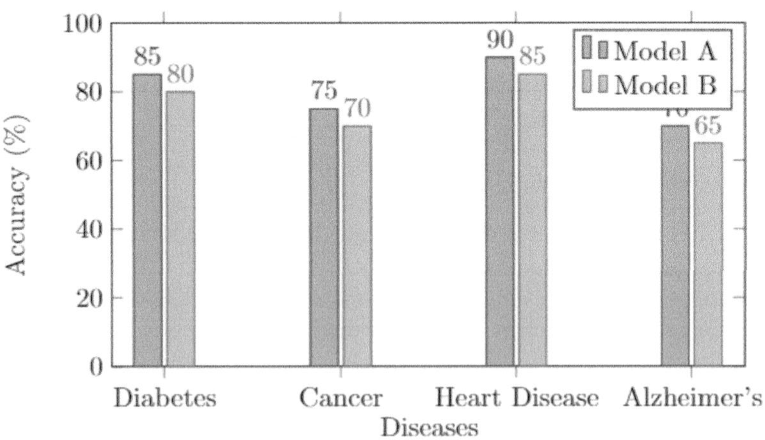

Figure 1. Comparative analysis of predictive models in various diseases.

4. Deep Learning in Disease Prediction

The field of Smart Systems for Health Care has seen Deep Learning (DL) become a major power in predicting diseases. Predictive analytics has advanced in previously unimaginable ways, thanks to deep learning techniques. They are able to automatically extract intricate patterns from extremely large datasets. The strength of profound learning plan structures make it great at foreseeing infections. Convolutional Neural Networks (CNNs), Recurrent Neural Networks (RNNs), and their hybrids are exceptionally adept at handling various kinds of medical data, including pictures, sequences, and structured clinical information (Rajesh and Rajendran 2022).

In imaging and pathology, DL is very good at making disease predictions based on images. As shown by plans like ResNet and DenseNet, Profound Convolutional Brain Organizations (DCNNs) have the best outcomes in errands like tracking down growths and arranging them. Gated Recurrent Units (GRUs) and Long Short-Term Memory (LSTM) networks have been extremely beneficial for chronic diseases like diabetes and heart problems. These designs consider the recording of time subtleties in persistent wellbeing records (Rahul and Monika 2019).

DL is sufficiently adaptable to consolidate information from various sorts of sources. Assembling information from various sources, similar to genomes, electronic wellbeing records, and shrewd sensors, makes it more straightforward to get a full image of an individual's wellbeing. Transfer learning is a crucial concept in deep learning, as it enables the application of knowledge gained from one task to another, improving efficiency and reducing the need for extensive retraining. This approach significantly enhances the training of disease prediction models by utilizing models that have already been trained on large datasets.

Since DL models are 'secret elements', the study centers around making them more straightforward to make sense of them and comprehend them. Methods like consideration cycles and saliency maps assist us with understanding how models decide, which is vital for getting specialists to trust them. When using DL to predict diseases, problems can arise when samples are not fair. Cost-sensitive learning and data addition are two important methods for reducing errors and making sure that models work well (Afiqah et al. 2020).

DL works even better when results from multiple models are combined using ensemble learning methods. Techniques like bagging and boosting help make models more reliable and useful in more situations. Edge computing has made it easier to predict diseases in real-time at the point of care. Putting lightweight DL models on edge devices lowers delay and makes health systems more flexible. Worries about tolerant security are raised by DL's utilization of enormous datasets. United learning and homomorphic encryption are seeming to be great choices since they let you train models without uncovering incorporated information.

Situations that require constant monitoring make use of DL's capacity for change. Versatile profound learning models, which are worked on over the long haul with new persistent information, can foresee illnesses in a way that is both dynamic and explicit. There are cutoff points to DL in regions where getting large data is difficult, as in uncommon sicknesses. Move learning and semi-managed learning are two strategies that help a great deal when there isn't an adequate amount of information.

While utilizing DL to anticipate sicknesses, moral issues like reasonableness, obligation, and inclination should be contemplated. Specialists are endeavoring to ensure that structures are utilized in a moral manner and that unforeseen impacts are kept to a base (Pokkunuri 2021).

DL models probably won't function admirably with loud information, which could make them less dependable. To make models more impervious to uproarious info, strong preparation strategies like dropout and ill-disposed preparing are utilized. DL and medical care laborers must cooperate. Human-computer based intelligence organization models, in which specialists give subject information, ensure that DL expectations are utilized capably in clinical navigation. When it comes to predicting diseases, DL is having trouble with the regulations in the healthcare industry. To safeguard patient data and privacy, regulations like the Health Insurance Portability and Accountability Act (HIPAA) must be adhered to (García et al. 2023).

With more review being finished in support learning and independent learning, the fate of profound learning in foreseeing illnesses looks brilliant. Prediction accuracy and model comprehension will both be enhanced by investigating new deep learning structures and approaches. It is vital for scientists, medical care laborers, and officials to cooperate universally. Strong DL models that can be used to predict diseases anywhere in the world are easier to create when information is shared and teams work together.

Algorithm 1. Deep learning algorithm for disease prediction.

```
Data: Input data: Features (X), Labels (Y)
Result: Trained deep learning model
Initialization of neural network architecture;
Initialize weights and biases;
Initialize hyperparameters (learning rate, batch size, etc.);
while not converged do
    Sample mini-batch from training data;
    Forward pass: Compute predictions using current model parameters;
    Compute loss between predictions and true labels;
    Backward pass: Compute gradients of loss with respect to model
      parameters;
    Update model parameters using gradient descent;
```

4.1 Neural Networks and Their Role in Disease Forecasting

The shape and function of the human brain has an impact on neural networks, which are becoming more crucial in disease prediction. These PC models, which are comprised of neurons-like connected hubs, are truly adept at gaining confounded designs from extremely enormous datasets. This makes them exceptionally helpful in the medical care field. As brain organizations can find complex associations in clinical information, they can likewise be utilized to distinguish various sicknesses. This gives us valuable data for early identification and visualization.

Intermittent brain organizations (RNNs) are vital with regards to anticipating infections. Their capacity to find transient connections in consecutive information makes it more straightforward to make forecasts about sicknesses that change after some time, as irresistible illnesses or long-haul conditions. By resolving the disappearing gradient problem, Long Short-Term Memory (LSTM) networks, a specific kind of RNN, enhance disease prediction models. This ensures that organizations can gain from long haul patient information in a dependable manner.

New versions of brain network designs, especially convolutional brain networks (CNNs), have completely changed how clinical image analysis is used to predict illnesses. The ability of CNNs to extract hierarchical features from medical image datasets enables them to precisely identify abnormalities that may indicate various diseases, including brain disorders and tumors. When using neural networks for disease prediction, it remains essential to consider how easily these models can be interpreted and understood. The goal of new developments in explainable artificial intelligence (XAI) is to make the decision-making process of neural networks less mysterious and provide doctors with clear views into model estimates (Madhumita and Parija 2009).

4.2 Case Studies and Advancements in Deep Learning for Disease Prediction

Real examples of how deep learning can be used to predict diseases are very exciting. In cardiology, for example, deep learning models examine electrocardiograms (ECGs) for signs of a heart condition. These models are much more accurate than the standard methods. Breakthroughs like these showcase the wonder of how deep learning can change diagnostics as we know it.

Using deep learning with varied sets of data sources enhances the ability to predict diseases. Integrating clinical data with genomics, proteomics, and other omics datasets expands our understanding of a subject in science. A vast collection of information enables more precise and fine-grained disease predictions. This initiative to unite experts from diverse backgrounds can drive advancements in precision medicine, fostering more accurate and personalized healthcare solutions.

Deep learning models are swift in adapting by transforming themselves so that newly evolved behaviors can be detected in viral diseases. For example, during global pandemics, recurring neural networks analyze diverse data sets like epidemiological datasets, clinical and environmental information. The analysis helps in early detection, surveillance, and reduction techniques.

These capabilities can be interesting in health-related natural language processing as more advanced deep learning frameworks, such as BERT (Bidirectional Encoder Representations from Transformers), improve. These systems assist in identifying inconsistencies in clinical notes and other medical data, improving the understanding of disease trends and enhancing diagnostic accuracy.

Algorithm 2. Neural network for disease forecasting.

```
Input: Historical health data, including patient demographics, medical history, and disease outcomes.
Output: Disease forecast for future time periods.
Initialize: Neural network architecture, hyperparameters, and training dataset.
Train Neural Network:
for each epoch do
    for each mini-batch in training dataset do
        Forward pass: Compute predictions using current model parameters.
        Compute loss: Calculate the difference between predictions and actual outcomes.
        Backward pass: Compute gradients of the loss with respect to model parameters.
        Update parameters: Adjust model parameters using optimization algorithm (e.g., stochastic gradient descent).
    end for
end for
Forecasting:
Input: Current health data for forecasting period.
Output: Disease forecast for the forecasting period.
Use trained neural network to make predictions on new data.
Evaluate Performance:
Assess the accuracy of disease forecasts using evaluation metrics such as accuracy, precision, recall, and F1-score.
Fine-tuning and Optimization:
Adjust hyperparameters, neural network architecture, or training process based on performance evaluation.
Repeat training and evaluation process until satisfactory performance is achieved.
Output: Trained neural network model for disease forecasting.
```

5. Genetic Algorithms for Disease Prediction

Intelligent health systems use genetic algorithms (GAs) for the identification of diseases. GAs have been widely used because they provide a robust approach to solving complex problems. These natural selection-based algorithms allow you to explore large solution spaces, which is critical for finding complex patterns in genetic data.

For morbidity prediction, GAs can handle large genetic datasets and provide a convenient approach for selecting features which would result in improved model accuracy. The power of the program to transform and expand helps locate genetic markers associated with different diseases. A major application of GAs in disease prediction is to improve the accuracy of AI models such as support vector machines and neural networks using parameter tuning. The approach with AI tools has been effective in addressing challenges arising healthcare challenges.

Consolidating hereditary calculations with highlight choice strategies makes it simpler to find hereditary elements that make individuals bound to get infections. By further developing responses again and again, GAs track down the best arrangement of hereditary elements that can be utilized. This makes infection forecast models more clear and more precise. Hereditary calculations can be utilized for something other than anticipating sicknesses in people. They can be utilized in populace-based examinations to find hereditary variations that are connected to higher sickness rates.

This gander at sicknesses according to the perspective of an entire local area assists us with understanding the hereditary scene of diseases better.

Utilizing GAs to work on the preparation of hereditary populace studies is an area of progressing research that plans to ensure that hereditary information is gathered and utilized productively. This app helps us understand how genetics and disease interact in complex ways and strengthens the statistics of studies. GAs are an adaptable method for managing the unequal datasets that are normal in medical care. They ensure that sickness expectation models function admirably across an extensive variety of populace gatherings. Since it tends to be changed, the program functions admirably, in actual medical services circumstances.

GAs can likewise be utilized to tune boundaries in customized medication, making illness conjecture models fit the extraordinary qualities of every patient. This customized strategy shows potential for accuracy medication and makes the way for additional particular medicines. As of late, hereditary calculations and other shrewd strategies, similar to profound learning, have begun cooperating to make blended models that utilize the most amazing aspects of both. The objective of this sort of blend is to get around the issues with individual calculations and work on the precision of forecasts in all.

Despite the fact that they work, GAs experience difficulty in answering changing genomic settings. A ton of work is being placed into creating programs that can change on the fly in light of changes in hereditary information over the long run. At the point when GAs are utilized in medical services, morals are vital, particularly with regards to ensuring that hereditary data is utilized dependably. There is still a challenge in achieving the ideal balance between protecting patients' privacy and making predictions that are as accurate as possible.

Hereditary calculation driven models are difficult to use in clinical settings since they are difficult to comprehend. To fix this, specialists are endeavoring to make computer-based intelligence strategies that can be clarified for make sickness expectation models more understood. Equal PC enhancements have made it simpler for GAs to be increased, which lets huge genomic datasets be examined rapidly. This adaptability is necessary for handling the increasing amount of genetic information.

In sickness forecast applications, where datasets are often brimming with mistakes and missing qualities, the way that GAs can deal with clamor or deficient hereditary information is a major upside. This makes the program more solid in clinical circumstances. A growing number of people are adding GAs to multi-omics data, combining genetic and other biological data. This interdisciplinary strategy aims to get a full image of the confounded connections that lead to illness.

GAs are additionally adaptable in anticipating illnesses after some time. This is because the program can adapt to changes in genetic data over time. This part of time is vital for understanding the number of sicknesses that change over the long haul. To enable the use of GAs in healthcare settings, efforts are still being made to develop platforms and tools that are simple to use for individuals unfamiliar with computers. This democratization aims to make advanced disease forecasting based on DNA algorithms more accessible to medical professionals.

Algorithm 3. Genetic algorithm for disease prediction.

```
Initialize population with random chromosomes
Evaluate fitness of each chromosome using disease prediction model
while not termination condition met do
    Select parents for mating based on fitness (e.g., tournament selection)
    Perform crossover to create offspring (e.g., one-point or two-point crossover)
    Perform mutation on offspring to introduce diversity (e.g., flip a bit with small probability)
    Evaluate fitness of new offspring
    Select individuals for the next generation (e.g., elitism or survivor selection)
end while
return Best chromosome as the predicted disease risk profile
```

Because genetic algorithms used to identify diseases are constantly changing, they need to be checked and compared with other methods frequently. The algorithmic outcomes are robust and can be applied across various datasets and networks due to rigorous testing. Incorporating feedback loops into genetic algorithms enhances their adaptability, allowing the model to refine itself based on the effectiveness of previous predictions. This capacity for self-advancement assists the program to improve itself continuously.

5.1 Genetic Algorithm Principles in Healthcare

In medical care, hereditary calculations (GAs) depend on the thoughts of developmental registering, which depend on the course of regular choice. In this model, GAs work by slowly changing a gathering of potential solutions to an issue, like how the 'natural selection' process works. Hybrid and change of potential arrangements are essential for this iterative interaction. They make a differed bunch that gradually merges on ideal or almost ideal arrangements. GAs are used in many areas of healthcare, such as predicting diseases, analyzing medical images, and making healthcare processes run more smoothly. When it comes to predicting diseases, GAs provide a strong computing structure that can handle the complexity of genetic data. These methods use the ideas of genetic evolution to make predictive models better. This lets us get useful patterns and connections from very big genomic datasets.

To use GAs to their full potential in healthcare, it is important to understand how they work at their core. To use GAs successfully to solve the complex problems in the healthcare field, both researchers and practitioners need to understand the differences between population evolution, crossover, and mutation operators. In the future, when the field grows, understanding these ideas better will help make genetic code uses in medicine better and more advanced.

5.2 Utilizing Genetic Algorithms for Disease Risk Assessment

Genetic algorithms are useful for figuring out how likely a disease is to occur because they give us a way to look at complicated genetic information in a structured way. When doing disease risk assessment, the goal is to find people whose genes make them more likely to get certain diseases. GAs are very good at this because they choose the best genetic markers and improve prediction models to make them more accurate.

Choosing the best feature groups from large genome datasets is a common way that genetic algorithms are used to figure out the risk of a disease. GAs go through a huge number of possible feature pairs over and over again until they find a group that gives the best predictions. This feature selection process not only make predictive models work better; it also helps find medically important genetic markers linked to higher disease risk.

Hereditary calculations are utilized for something other than anticipating a singular's possibility of getting a sickness. They are likewise utilized for concentrates on that gander at different people in a gathering. This is on the grounds that GAs assist with finding hereditary variations that are connected to higher sickness rates in specific gatherings by taking a gander at enormous datasets. This populace-focused-view assists us with studying what hereditary qualities means for the general weight of illnesses, which makes it workable for more engaged general wellbeing activities.

All in all, utilizing hereditary calculations to sort out the gamble of getting a sickness demonstrates the way that they can work on the precision of estimating models and make the most common way of picking highlights more productive. These algorithms are getting better and better, and by figuring out the complicated connection between genes and disease risk, they could be of great assistance to efforts to improve population health and personalized medicine.

6. Ensemble Methods in Disease Prediction

Group strategies have become exceptionally helpful for anticipating sicknesses since they utilize the consolidated information on many models to make expectations more exact and solid. Individual models probably won't have the option to catch the intricacy of medical care information completely. Group techniques, then again, consolidate various speculations to make a complete and more precise forecast. This part discusses the thoughts behind outfit procedures and how they can be utilized to foresee illnesses. It additionally discusses how effective they are and the way that they could assist with further developing medical services results.

6.1 Understanding Ensemble Techniques in Disease Forecasting

While attempting to foresee infections, gathering techniques join a few expectation models, each educated on an alternate subset of information or with an alternate calculation. The basic idea is to make a better estimate than any one model by taking advantage of the differences between models. Gathering procedures are truly

Algorithm 4. Ensemble methods in disease prediction.

```
Input: Dataset D, Base learners B, Ensemble method E
Output: Ensemble model M

function TrainEnsemble(D, B, E)
    M ← ∅                                         ▷ Initialize ensemble model
    for each base learner b in B do
        Train base learner b on dataset D
        M ← M ∪ {b}                               ▷ Add base learner to ensemble model
    end for
    return M
end function

function Predict(M, x)
    predictions ← ∅
    for each base learner b in M do
        prediction ← Predict using b on input x
        predictions ← predictions ∪ {prediction}
    end for
    return E(predictions)   ▷ Combine predictions using ensemble method E
end function
```

adaptable, and techniques like stowing, helping, and stacking show this. Every one adds its own assets to the complete prescient structure.

Gathering strategies consider the vulnerabilities and challenges that accompany medical services datasets. They offer areas of strength to decrease model difference and inclination. Gathering techniques give a steadier and more precise image of the patterns in sickness information by consolidating the outcomes from various models. People who want to use ensemble methods effectively in real-world healthcare situations must be aware of the intricate interactions between them.

6.2 Integrating Ensemble Methods for Enhanced Disease Prediction Accuracy

It looks like adding group strategies to models that anticipate illnesses could make them significantly more exact and solid. Issues like overfitting and speculation botches are handily fixed with troupe techniques, yet they are difficult to fix with standard single-model methodologies. By blending the most awesome aspects of various models, outfit techniques make it simpler for the model to function admirably with new information, which is vital in the continually changing field of illness expectation.

Additionally, ensemble methods make it simpler to utilize a wide range of data, including clinical, environmental, and genetic data. This mix makes it conceivable to get a full image of the many variables that influence sicknesses. When utilized accurately, troupe techniques make it conceivable to make prescient models that are right as well as adaptable enough to manage the various kinds of information that are tracked down in medical care.

7. Comparative Analysis and Best Practices

To figure out how well smart techniques work at foreseeing sicknesses, we want to do a full correlation of execution measurements that will assist us with involving them in various medical care circumstances. In this section, the presentation measures associated with canny strategies are taken a gander at exhaustively, alongside the upsides and downsides of utilizing these measurements to pass judgment on prescient models. It gives helpful data about the most effective ways to involve shrewd methods for various diseases, with an emphasis on a customized approach for the best outcomes.

7.1 Evaluating Performance Metrics of Intelligent Techniques

A profound comprehension of progress measures is required to decide how well sharp strategies work at foreseeing infections accurately. The model's ability to recognize positive and negative situations is evaluated using important metrics like sensitivity, specificity, accuracy, recall, and F1 score. This section extends beyond conventional metrics by examining the Area Under the Receiver Operating Characteristic Curve (AUC-ROC) and the Precision-Recall Curve. These evaluation frameworks provide a comprehensive assessment, addressing challenges posed by imbalanced datasets and varying disease prevalence rates.

Since illnesses are unique, it is essential to investigate space-explicit measures as a component of the correlation study. For instance, speed and early revelation might be vital regarding the irresistible sickness gauge, implying that standard estimates should be rethought. This section aims to provide researchers and practitioners with the information they require to effectively select and comprehend performance measures. This will consider a more precise and valuable assessment of shrewd techniques in sickness expectation.

Figure 2 shows how well smart techniques work at foreseeing infections by taking a gander at their presentation measures and giving a visual portrayal of this. By estimating these strategies with various measurements, similar to exactness, accuracy, memory, and F1 score, among others, it gives a full image of their upsides and downsides. This image shows that smart methods work better compared to old ones, and it demonstrates the way that they could change how sicknesses are anticipated by making them more precise and quicker. It additionally postulates that it is essential to continue testing and working on these techniques to make them far superior at anticipating the future and make them work with the changing idea of medical care information.

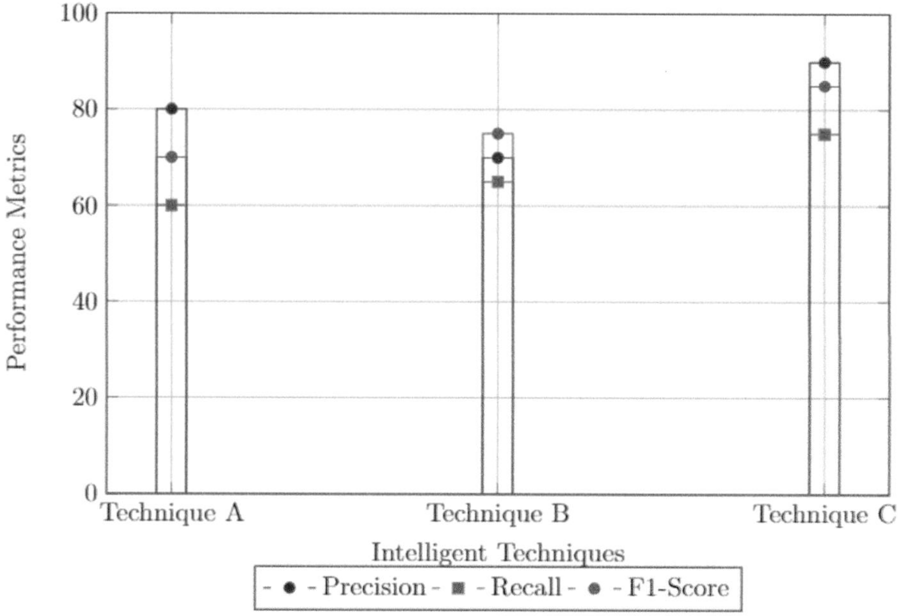

Figure 2. Evaluating performance metrics of intelligent techniques.

7.2 Recommendations and Best Approaches for Various Diseases

For the best anticipated results, astute strategies should be customized to the particulars of every illness. This part gives ideas and the most effective ways to involve shrewd techniques in various medical services settings. With regards to constant infections, long haul gauges and hazard delineation might be more significant, and that implies that ceaseless information and individual patient profiles should be utilized. At the point when there is a spreading sickness, thinking that it is early and keeping it contained turns out to be more significant, which calls for ongoing information coordination and dynamic model change.

The part additionally discusses the significance of having the option to portray astute techniques, particularly while moving from study to clinical use. Straightforward models assist with building trust among specialists and patients, which is vital for integrating savvy strategies into ordinary clinical practices. There are helpful suggestions on how to find the right balance between model complexity and readability so that clever methods can be used easily in all healthcare settings.

8. Conclusion

Since clever methods were added, the field of disease diagnosis has changed a lot. This section took care of a ton of ground about man-made brainpower, AI, and PC strategies, showing how significant they are in having an impact on how medical services are finished. This study explored various intelligent techniques, including genetic algorithms and ensemble methods, each contributing to greater accuracy and effectiveness in disease prediction models. Unlike traditional approaches, intelligent techniques are better equipped to handle large and complex healthcare datasets, enabling more precise and scalable predictive analysis. Intelligent techniques are capable of locating connections and patterns that other methods miss. Joining complex PC programs with clinical information helps find infections early and makes care more customized and adaptable for every patient. Enlivened by regular choice, hereditary calculations find their direction through the muddled hereditary scene, making forecast models more exact by further developing them. Ensemble methods, on the other hand, use the combined intelligence of several models. This is a strong and flexible way to deal with the complexity of healthcare data. The need for a complex and situation-specific approach is made clear by critically examining success measures and outlining best practices. Transparency and interpretability become vital as intelligent techniques move from research projects to clinical uses. They help healthcare professionals trust them and ensure they work well with regular medical practices. The coming together of innovative techniques and disease forecasts mark the start of a new healthcare era marked by proactive, accurate, and patient-centered methods. As these methods are improved and used more, they could change not only predictive models but also how healthcare is delivered, leading to better results for patients and better health for the whole community.

References

Afiqah Mohd Johari, N. A., Mohamad, N. and Isa, N. (2020). Smart self-checkup for early disease prediction. 2020 IEEE International Conference on Automatic Control and Intelligent Systems (I2CACIS). Shah Alam, Malaysia, pp. 33–38, doi: 10.1109/I2CACIS49202.2020.9140205.

Amin, Syed, Agarwal, Kavita and Beg, Rizwan. (2013). Genetic neural network-based data mining in prediction of heart disease using risk factors. 2013 IEEE Conference on Information and Communication Technologies, ICT 2013. 1227–1231. 10.1109/CICT.2013.6558288.

Babu, M. S. and Karthick, V. (2022). Improving accuracy in intelligent coronary heart disease diagnosis prediction model using support vector clustering technique compared over random forest classifier algorithm. 2022 International Conference on Innovative Computing, Intelligent Communication and Smart Electrical Systems (ICSES).

Choudhary, C., Nagra, L. S., Das, P., Singh, J. and Jamwal, S. S. (2023). Optimized ensemble machine learning model for chronic kidney disease prediction. 2023 International Conference on Computing, Communication, and Intelligent Systems (ICCCIS). Greater Noida, India, pp. 292–297, doi: 10.1109/ICCCIS60361.2023.10425073.

Djerioui, M., Brik, Y., Ladjal, M. and Attallah, B. (2020). Heart Disease prediction using MLP and LSTM models. 2020 International Conference on Electrical Engineering (ICEE). Istanbul, Turkey, pp. 1–5, doi: 10.1109/ICEE49691.2020.9249935.

Franklin, R. G. and Muthukumar, B. (2020). Survey of heart disease prediction and identification using machine learning approaches. 2020 3rd International Conference on Intelligent Sustainable Systems (ICISS). Thoothukudi, India, pp. 553–557, doi: 10.1109/ICISS49785.2020.9316119.

García-Ordás, M. T., Bayón-Gutiérrez, M., Benavides, C. et al. (2023). Heart disease risk prediction using deep learning techniques with feature augmentation. Multimed Tools Appl 82: 31759–31773. https://doi.org/10.1007/s11042-023-14817-z.

Geeitha., S., Karthikeyan., P., Aravinth., S., Nachiappan., P. and Jagadeesh, S. (2023). Contagious disease prediction using random forest algorithm interpolated with fuzzy model. 2023 International Conference on Sustainable Computing and Data Communication Systems (ICSCDS). Erode, India, pp. 674–679, doi: 10.1109/ICSCDS56580.2023.10104962.

Ghazal, T. M., Rehman, A. U., Saleem, M., Ahmad, M., Ahmad, S. and Mehmood, F. (2022). Intelligent model to predict early liver disease using machine learning technique. 2022 International Conference on Business Analytics for Technology and Security (ICBATS). Dubai, United Arab Emirates, pp. 1–5, doi: 10.1109/ICBATS54253.2022.9758929.

KS. K., Kousar, N. and J, M. (2020). Coronary artery disease prediction using data mining techniques. 2020 3rd International Conference on Intelligent Sustainable Systems (ICISS).

Lafta, R., Zhang, J., Tao, X., Li, Y. and Tseng, V. S. (2015). An intelligent recommender system based on short-term risk prediction for heart disease patients. 2015 IEEE/WIC/ACM International Conference on Web Intelligence and Intelligent Agent Technology (WI-IAT). Singapore, pp. 102–105, doi: 10.1109/WI-IAT.2015.47.

Lakshmanarao, A., Srisaila, A. and Kiran, T. S. R. (2021). Heart disease prediction using feature selection and ensemble learning techniques. 2021 Third International Conference on Intelligent Communication Technologies and Virtual Mobile Networks (ICICV), pp. 994–998.

Madhumita Pal and Smita Parija. (2021). J. Phys.: Conf. Ser. 1817: 012009.

Pokkunuri, D. (2021). Heart disease prediction in medical industries using data mining techniques. International Journal for Research in Applied Science and Engineering Technology 9(3): 1289–1291. https://doi.org/10.22214/ijraset.2021.33518.

Priyanga, P. and Naveen, N. C. (2017). Web analytics support system for prediction of heart disease using naive Bayes weighted approach (NBwa). 2017 Asia Modelling Symposium (AMS). Kota Kinabalu, Malaysia, pp. 21–26, doi: 10.1109/AMS.2017.12.

Rahul, Bansal, H. and Monika. (2019). Classification techniques used in sentiment analysis & prediction of heart disease using data mining techniques: Review. 2019 International Conference on Issues and Challenges in Intelligent Computing Techniques (ICICT). Ghaziabad, India, pp. 1–6, doi: 10.1109/ICICT46931.2019.8977707.

Raja, A. A., Guftar, M., Irfan-ul-Haq, Khan, T. A. and Greibl, D. (2016). Intelligent Syncope Disease prediction framework using DM-ensemble techniques. 2016 Future Technologies Conference (FTC). San Francisco, CA, USA, pp. 269–273, doi: 10.1109/FTC.2016.7821621.

Rajesh, T. R. and Rajendran, S. (2022). Intelligent multi-agent reinforcement learning based disease prediction and treatment recommendation model. 2022 International Conference on Augmented Intelligence and Sustainable Systems (ICAISS). Trichy, India, pp. 216–221, doi: 10.1109/ICAISS55157.2022.10010747.

Rane, A. L. (2018). A survey on intelligent data mining techniques used in heart disease prediction. 2018 3rd International Conference on Computational Systems and Information Technology for Sustainable Solutions (CSITSS).

Sharma, S. and Singhal, A. (2023). A novel heart disease prediction system using XGBoost classifier coupled with ADASYN SMOTE. 2023 International Conference on Computing, Communication, and Intelligent Systems (ICCCIS). Greater Noida, India, pp. 76–81, doi: 10.1109/ICCCIS60361.2023.10425095.

Singh, Archana and Kaswan Kuldeep Singh. (2024). Improved methodology for breast cancer prediction through integration of hard voting ensemble classifier on WDBC data set. Communications on Applied Nonlinear Analysis 31(6s): 179–192.

Sowmiya, C. and Sumitra, P. (2017). Analytical study of heart disease diagnosis using classification techniques. 2017 IEEE International Conference on Intelligent Techniques in Control, Optimization and Signal Processing (INCOS). India, pp. 1–5, doi: 10.1109/ITCOSP.2017.8303115.

Vij, A. and Kaswan, K. S. (2023, January). Prediction of lung cancer using convolution neural networks. In 2023 International Conference on Artificial Intelligence and Smart Communication (AISC) (pp. 737–741). IEEE.

3

Role of Image Processing in Disease Prediction and Early Diagnosis

1. Introduction

Based on the concepts of genetics and natural selection, genetic algorithms (GAs) are powerful ways of improving things. Since they can rapidly glance through enormous, complex element spaces, hereditary calculations are turning out to be more well known in the field of illness expectation (Kumar et al. 2024).

Hereditary calculations are truly adept at making highlight determination strategies more productive. This assists them with taking care of the issue of getting the most valuable highlights from extremely enormous arrangements of clinical pictures. Through a course of determination, hybrid, and development, GAs work on the capacity of prescient models to differentiate between things by finding characteristics connected to illness appearance.

When GAs are utilized, the degree to which disease forecast models can be generalized is markedly different. GAs play a crucial role in developing robust and adaptable prediction models by minimizing overfitting and selecting features that capture the underlying patterns of diseases. GAs can be utilized for something other than picking highlights; they can be utilized to make arrangement models better over the long haul. Genetic algorithms improve over time, taking into account the various types of diseases and their levels of complexity, resulting in extremely accurate models (He et al. 2023).

Hereditary calculations are generally excellent at multi-objective enhancement, which is vital for illness estimate since it necessities checking out of a number of variables simultaneously. It is feasible to look at responsiveness, particularity, and other execution measures with this element and forecast models can be made. GAs let you gain from information in an adaptable manner by changing model boundaries as conditions change. This adaptability is exceptionally useful while attempting to

DOI: 10.1201/9781032722344-3

think about what illnesses individuals will get, on the grounds that as sicknesses change or new sorts show up, models need to adjust over the long haul.

In clinical imaging, it is typical for measurements to be uneven, with additional sound cases than debilitated ones. Hereditary calculations are an effective method for fixing this issue since they even out the dispersion of classes during model preparation. This makes it doubtful that slanted datasets will influence how well the model predicts. Hereditary calculations are truly adept at tracking down complex associations among characteristics and disposing of copies. This component ensures that illness forecast models incorporate significant connections while disposing of pointless or copy information. This produces results that are more obvious and more exact (Jiao et al. 2024).

Genetic algorithms can quickly process large datasets and are highly scalable due to their parallel operation. Parallelizing genetic algorithms accelerates model development and convergence in disease prediction, where datasets are frequently quite large. Infection expectation models work stunningly better when they utilize half and half techniques that join hereditary calculations with other PC strategies. At the point when you join GAs with AI strategies or profound learning structures, the most amazing aspects of each work together to improve the framework.

Hereditary calculations continuously assist with anticipating sicknesses by rapidly working on model boundaries. This capacity to change progressively is vital in clinical circumstances, where speedy choices are expected to help patients. GA excel at handling noise and incomplete medical image data due to their strong foundation in natural selection principles. GAs work on the precision of illness conjecture models in many information circumstances by thinking of answers in any event, when there are questions.

By adjusting attributes to every patient's experience, hereditary calculations make it simpler to make custom models for anticipating infections. This customized technique makes diagnosis more exact by considering how illnesses appear distinctively in changed gatherings of patients. Since DNA recipes are clear, infection forecast models are more obvious. Clinicians can discover the elements that influence assumptions, which assembles trust in the model's dynamic cycle.

Genetic Algorithms helps in disease prediction by adapting pre-trained models to the new unknown datasets for the models. This helps to enhance the ability of models in generalization and improve the forecasting and model performance. This is vital for utilizing what you realize in one area of clinical imaging and working on your capacity to recognize what will occur in different associated regions. The system is guaranteed to be able to adapt to shifting disease patterns by employing genetic algorithms to dynamically adjust model parameters. This ability is particularly valuable for sicknesses whose side effects or recurrence change in the long run.

Genetic algorithms make evolutionary ensemble methods, which allow multiple models to change simultaneously, possible. This strategy utilizes different methods to make forecasts more precise in sickness cases with a ton of variety and intricacy. Genetic algorithms are very helpful for dealing with problems that have a lot of dimensions in the age of 'big data' in medical imaging. For making solid and calculably attainable sickness figure models, they should have the option to travel through enormous component regions rapidly (Jomeiri et al. 2024).

Hereditary calculations can likewise assist with social issues that surface while attempting to foresee illnesses. As GAs work to make models all the fairer and more liberated from predisposition, they assist with making more exact and fair expectation instruments. Hereditary calculations have been truly adept at anticipating illnesses, yet more review is expected to tackle issues like how to comprehend them, how to make them work on bigger datasets, and how to ensure that they are solid. To completely involve hereditary calculations for illness identification later on, work should be finished to work on these techniques and empower cooperation between PC researchers and clinical experts.

1.1 Image Processing Techniques in Disease Prediction

In the ebb and flow of medical services, utilizing picture handling strategies to recognize illnesses is turning out to be increasingly significant. With the improvement of further developed clinical imaging techniques, getting helpful information from huge datasets has turned into a significant piece of prescient displaying.

As picture handling has filled in medical care, it has denoted a change from subjective to quantitative survey. Computational techniques are utilized to track down complex examples and peculiarities in clinical pictures. These strategies incorporate picture division and component extraction. Because it separates and defines areas of interest in medical photographs, image segmentation is an essential step in disease prediction. This makes it more straightforward to get the right component extraction, which makes it conceivable to make solid forecast models.

A vital piece of transforming crude picture information into valuable structures is include extraction. Procedures like surface investigation and shape markers make it conceivable to quantify little contrasts, which give illness expectation models valuable elements. Data from different imaging techniques should be assembled to make exact sickness forecasts. Picture enlistment techniques unite information from various sources to get a full image of the sickness scene. Image processing for disease detection has evolved significantly with the integration of AI techniques, particularly deep learning models. It is remarkable how well recurrent models and convolutional neural networks (CNNs) can learn hierarchical structures from medical images.

In spite of a leap forward, obstacles stay in picture handling for illness finding, remembering changeability for imaging strategies, information assortment, and interpretability of confounded models. Getting these issues addressed is fundamental for prescient models to be utilized in clinical settings. Early assessment is a vital piece of good medical services, since it considers fast activity and improved results for patients. Utilizing picture handling in early identification is an extraordinary method for utilizing information driven plans to find early indications of sicknesses that are as yet gentle (Sathiyamoorthi et al. 2021).

The connection between imaging strategies and early discovery is vital for finding illness signs that have been covered up. To provide a more complete picture of a disease, multiple imaging techniques, such as magnetic resonance imaging (MRI), computed tomography (CT), and positron emission tomography (PET), collaborate. Radiomics is another region that spotlights on getting numeric data from clinical pictures. Radiomics works on the exactness of forecast models by estimating little surfaces and examples. This allows sicknesses to be viewed as ahead of schedule and precisely depicted.

When clinical and picture information are joined, it makes it simpler for models to make expectations. At the point when you consolidate clinical data with imaging results, you get a more complete image of the patient, which helps sickness forecast models work better. PC Helped Analysis (computer aided analysis) frameworks, which depend on picture handling, are exceptionally valuable for specialists. These frameworks, which are controlled by AI calculations, assist specialists with sorting out what clinical pictures mean, which further develops their appraisal abilities (Gao 2021).

Picture handling techniques make it conceivable to get precise picture signals. As estimated indications of changes in the body, these biomarkers assist with building a numeric and objective reason for illness conjecture. Utilizing picture handling to think about what infections individuals have could help decide on customized treatment. Utilizing imaging information to make prescient models that are well-defined for every patient's profile takes into consideration risk evaluation and intercession designs that are made only for that patient.

Picture hereditary qualities are another region that investigates what hereditary factors and picture information mean for one another. Disease prediction models become more accurate when genetic data and image processing techniques are combined, and this helps us understand how genetic factors contribute to diseases. Clinical Choice Emotionally supportive networks (CDSS) are worked around picture handling, which assists specialists and medical attendants with settling on brilliant decisions. CDSS works on the speed and exactness of illness expectation in clinical cycles by giving constant bits of knowledge and ideas (Prakash and Saradha 2023).

While utilizing pictures to anticipate sicknesses, pondering moral issues is vital. Concerns about patient privacy, data security, and the proper application of prediction models necessitate that image processing technologies be utilized with caution and morality. For estimate models to acquire trust and acknowledgment in the clinical local area, they should be straightforward and make sense. Making picture handling-based models clearer makes them more helpful for settling on medical services choices. Picture handling in illness identification has a splendid future in front of it as innovation continues. Probably the main patterns that will influence the scene are utilizing ongoing information, taking advantage of 3D pictures, and making things clearer.

2. Imaging Modalities in Disease Detection

As imaging modalities allow doctors to see inside the body and identify issues, imaging techniques are crucial for disease detection. From outdated X-beams to cutting edge useful imaging, this section takes a gander at the extensive variety of imaging methods that have assisted specialists with finding sicknesses all the more rapidly.

X-beam imaging in illness finding: Because it lets doctors see inside the body and find problems without hurting the patient, X-ray imaging is still an important part of medical diagnosis. It may be utilized to track down breaks and check for lung issues, giving specialists significant data about numerous infections (Göndöcs and Dörfler 2024).

For nitty gritty life structures: Figured Tomography (CT) utilizes X-beams to make cross-sectional photos of the body, which give a great deal of data about the body's construction. Since it can show organs, veins, and delicate tissues, CT is a significant device for tracking down sicknesses, particularly in oncology and cardiovascular medication. Attractive reverberation imaging (X-ray) and the partition of delicate tissues: Based on magnetic resonance, MRI is excellent at distinguishing between various types of soft tissue. Its high difference clearness helps track down issues in the mind, outer muscle framework, and stomach organs, which gives us data about numerous sicknesses.

Ultrasound imaging for constant representation: Utilizing sound waves, ultrasound imaging allows you to see objects inside the body progressively. Ultrasound is ordinarily utilized in pregnancy, heart, and gut imaging. It is a protected and dynamic method for finding sicknesses without utilizing ionizing radiation.

Atomic medication and sub-atomic imaging: Radioactive tracers are utilized in atomic medication to perceive how the body functions at the sub-atomic level. Methods like positron emanation tomography (PET) and single-photon discharge registered tomography (SPECT) assist with finding sicknesses by showing how digestion and capability change in various circumstances.

Mammography for tracking down bosom malignant growth: Mammography plays a crucial role in the detection and treatment of breast cancer. It is a particular X-beam technique. Advanced radiography and tomosynthesis have become more delicate, which assists with early discovery and brings down the demise rate from bosom malignant growth.

For mind action planning with practical X-ray (fMRI): Practical X-ray records changes in blood stream that show how the cerebrum is functioning. fMRI is involved a ton in neuroscience and nervous system science to help find and depict brain sicknesses by following the progressions in capability that accompany them.

DWI (weighted by diffusion): DWI (weighted by diffusion) in MRI oncology imaging identifies the erratic movement of water molecules within tissues. DWI is for the most part utilized in oncology to portray growths, check how well therapy is working, and track down early indications of disease.

Chest X-beams and lung diseases: Chest X-beams are one of the main ways of tracking down lung sicknesses. Chest X-beams assist with finding respiratory ailments almost immediately and watch out for them by showing everything from diseases to cellular breakdown in the lungs (Arain et al. 2023).

Useful imaging in cardiology: Atomic cardiology and cardiovascular X-ray are two sorts of practical imaging that give specialists significant data about how the heart functions and how much blood courses through it. These strategies assist specialists with sorting out what heart sicknesses an individual has and how to treat them so the patient has an improved outcome.

Imaging in the infrared and thermography: Imaging in the infrared or thermography utilizes warm light to show changes in temperature. Accordingly, thermography aids in the identification of skin issues, circulation issues, and inflammation diseases when utilized in dermatology and vascular medicine (Rahman et al. 2024).

Double Energy X-beam Absorptiometry (DXA) in bone wellbeing: DXA estimates bone mineral thickness, which measures the strength of the bones and the possibilities of breakage. Most of the time DXA is used to check for osteoporosis and helps find and treat spine illnesses early.

Photoacoustic imaging for better differentiation: This kind of imaging utilizes both conventional imaging strategies and laser-actuated ultrasound to improve contrast in delicate tissues. At the point when utilized in malignant growth and vascular imaging, photoacoustic techniques help find and portray sicknesses from the beginning.

Cineangiography and interventional radiology: Interventional radiology and cineangiography are frequently utilized together on the grounds that they let specialists see blood veins and systems progressively. This technique has a major effect in finding sicknesses since it assists specialists with diagnosing circulatory diseases and plan therapies.

Utilization of endoscopy for direct visual assessment: Endoscopic imaging allows you to see organs and designs inside the body straightforwardly. In gastroenterology, urology, and pulmonary medicine, endoscopy is utilized to assist in the early detection of wounds, tumors, and inflammatory conditions.

Retinal imaging using Optical Coherence Tomography (OCT): OCT takes high-goal cross-sectional pictures utilizing low-cognizance light. OCT is frequently utilized in ophthalmology to aid in the early detection and monitoring of retinal diseases such as diabetic retinopathy and macular degeneration.

Enhancing separation through spectral imaging: Otherworldly imaging gathers information at various frequencies, which makes it simpler to differentiate between cells. Ghostly imaging is utilized in malignant growth and pathology to assist with portraying cancers and track down early indications of sickness (Jung et al. 2021).

Cryo-EM, or cryo-electron microscopy: Cryo-EM is a better type of magnifying lens that allows you to see living designs plainly. According to cryo-electron microscopy (Cryo-EM) is used in structural biology and drug development to provide atomic-level understanding of molecular structures and diseases.

In Figure 1, different imaging methods used to find sicknesses are shown. Every one gives an alternate perspective on the construction and capability of the human body. Conventional techniques, for example, X-beams and ultrasound, produce two-layered pictures that are exceptionally useful in recognizing many sicknesses, from broken unresolved issues. CT and X-ray examines produce three-layered pictures that give a more complete image of the body. These pictures are exceptionally

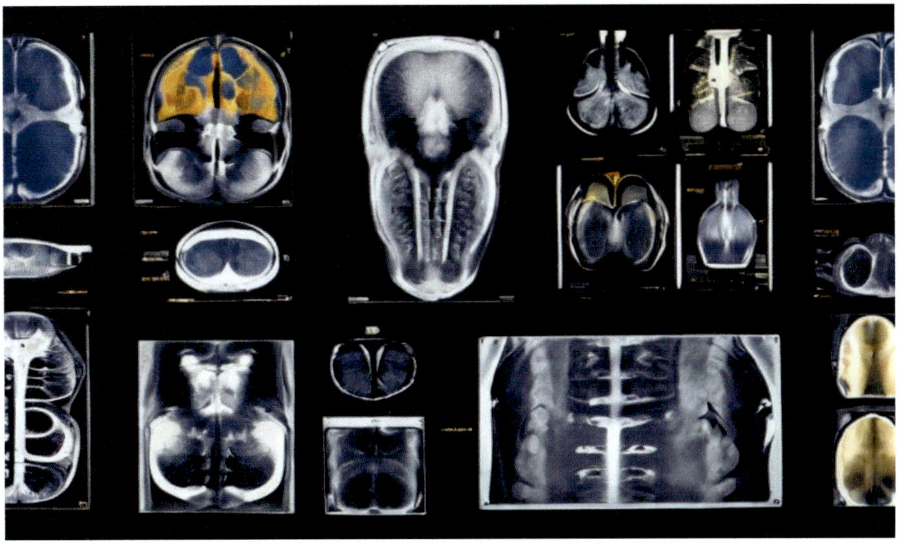

Figure 1. Imaging modalities in disease detection.

useful in finding and watching out for serious sicknesses. Positron Discharge Tomography (PET) filters, which are frequently utilized with CT or X-ray examines, show how digestion works and assist specialists with sorting out how sicknesses are advancing. Optical imaging techniques like Photoacoustic Imaging are more current enhancements. Ultrasound and laser-induced multispectral photoacoustic signals provide high-resolution images of molecules, metabolism, and functions. These various strategies, utilized alone or together, assist specialists with making more exact judgments, plan better medicines, and watch out for how sicknesses are advancing.

2.1 Comparative Analysis and Impact on Disease Detection and Prediction

At the point when you think about various brilliant frameworks in medical services, you can perceive what they mean for finding and anticipating sicknesses in various ways. Customary approaches to diagnosing frequently include glancing back at patient records and signs, which doesn't assist with making forecasts (Hu et al. 2023). Then again, brilliant frameworks driven by computerized reasoning (man-made intelligence) and AI (ML) can take a gander at exceptionally enormous and fluctuating datasets. This allows them to find sicknesses early and anticipate patterns that customary strategies could miss. These shrewd frameworks make it more straightforward to track down sicknesses, yet they additionally assist with making expectations by getting on little signs and chance variables. Allowing personalized, data-driven actions that improve patient outcomes alters the delivery of health care. Table 1, analysis of the tools for disease detection and prediction.

Table 1. Analysis of tools for diseases detection and prediction.

Technology/ Method	**Advantages**	**Limitations**	**Applications**	**Examples**
Machine Learning	- High accuracy in pattern recognition - Ability to analyze large datasets - Continuous learning and adaptation	• Requires substantial training data • Interpretability challenges • Susceptible to bias	- Image recognition in radiology - Predictive analytics in genomics	- Deep learning algorithms for cancer detection - Predictive models for diabetes risk
Genomic Sequencing	- Identification of genetic markers - Personalized medicine - Early detection of genetic diseases	• Costly and time consuming Ethical and privacy concerns • Limited understanding of some genetic variations	- Cancer genomics - Pharmacogenomics	- Human Genome Project - Cancer genomic profiling
IoT in Healthcare	- Real-time monitoring - Remote patient management - Data-driven insights	• Security and privacy risks • Integration challenges • Dependence on connectivity	- Chronic disease management - Wearable health devices	- Smartwatches for heart rate monitoring - Implantable sensors for diabetes management
Immunohisto-chemistry	- Tissue-specific analysis - Biomarker identification - Histopathological diagnosis	• Labor-intensive and time-consuming • Subject to variability in interpretation • Limited quantitative information	- Cancer diagnosis and prognosis - Infectious disease detection	- HER2 testing in breast cancer - CD markers for immune response
Telemedicine	- Increased accessibility to healthcare - Remote consultations - Reduced healthcare costs	• Limited physical examination • Technical barriers for some populations • Dependency on internet infrastructure	- Remote patient monitoring - Consultations for minor illnesses - Mental health support	

3. Suitability of Imaging Modalities for Disease Applications

Checking which imaging techniques turn out best for various diseases is a significant to ensure that conclusion strategies are custom-made to the points of interest of every sickness. Contextual investigations show that these techniques are considerably more valuable for diagnosing sicknesses. With the assistance of contextual analyses, the accompanying lines make sense of why certain imaging strategies are better for specific sicknesses:

X-Beam Imaging

- *Propriety*: X-beam imaging is particularly good for looking at bones and tracking down issues in the chest. It is much of the time used to figure out what's up with broken bones, disengaged joints, and lung issues.
- *Contextual analysis*: In the event of the possibility of a broken bone, X-beam imaging is a speedy and economical method of diagnosis. The subsequent x-beams show exhaustively the size and area of the break, which assists specialists with choosing the most effective way to treat it.

MRI: A Type of Magnetic Resonance Imaging

- *Fittingness*: X-ray is perfect for neuroscience, joint investigations, and circulatory tests since it doesn't utilize hurtful radiation and has an extraordinary difference for delicate tissues.
- *Contextual investigation*: X-ray is a vital instrument for diagnosing numerous scleroses. It makes demyelination visible in the brain and spinal cord, which aids in the early detection of issues and the planning of treatments.

CT (Computerized Tomography)

- *Suitability*: CT checks are frequently utilized for head, chest, tummy, and hip examinations since they produce exceptionally nitty gritty 3D pictures. It is helpful for tracking down skeletal imperfections.
- *Contextual investigation*: At the point when cellular breakdown in the lungs is being analyzed, CT filters make it more straightforward to find bronchial cancers and assist with sorting out how far along the sickness is. According to the National Lung Screening Trial Research Team (2011), this aids physicians in devising the most effective treatments for their patients.

Positron Discharge Tomography-Processed Tomography (PET-CT)

- *Fittingness*: PET-CT fusion is useful for studying cancer because it combines metabolic and structural data to identify and describe the location and type of the disease.
- *Contextual investigation*: PET-CT helps specialists assess the extent of cancer spread, guiding decisions on surgery or essential treatments. This integrated approach enhances diagnostic accuracy.

Molecular Imaging with Single Photon Discharge Figured Tomography (SPECT)

- *Propriety*: SPECT is a sort of atomic medication that is in many cases utilized in heart and nervous system science. It tends to be utilized for practical atomic imaging.
- *Contextual investigation*: During myocardial perfusion imaging, SPECT aids in determining how much blood is reaching the heart muscles. This helps track down coronary vein sickness and plan therapies.

Ultrasonography

- *Propriety*: Ultrasonography functions admirably for constant pictures, particularly in vascular examinations and pregnancies. It uses no harmful energy and does not harm the patient.
- *Contextual analysis*: During childbirth, ultrasound is crucial for monitoring the fetus's development and for identifying issues. The AIUM Practice Boundary for the Presentation of Obstetric Ultrasound (2018) gives helpful information for dealing with high-risk pregnancies.
- *Importance and Effect of Visual Data Catch*: X-beams, X-rays, and CT checks are vital devices in present day medication for finding illnesses since they show photos of the body. These imaging techniques give specialists clear pictures of body parts and changes that happen due to sickness, which assists them with pursuing right conclusions and shrewd therapy decisions.

Critical Job of X-Beams in Illness Recognizable Proof

- X-beams are, one of the main ways of tracking down illnesses, particularly in muscular health and chest radiography. X-beams assist specialists in seeing issues like cracks, separations, and lung sicknesses by allowing them to see the bones and joints.
- X-beams are vital in muscular health for finding broken bones and issues with joints so that the right treatment plans can be made.
- Chest X-beams assist with finding lung sicknesses like contaminations, growths, and different issues so therapy can start at the earliest opportunity.

Exploring the Complexities of X-ray in Illness ID

- X-ray is fundamental for the cerebrum, joint examinations, and circulatory tests since it can take high-goal pictures with extraordinary differentiation among hard and delicate tissues.
- In nervous system science, X-ray assists in finding neurological sicknesses with loving different sclerosis by showing bunches of insights regarding cerebrum designs and issues.
- Musculoskeletal MRI contributes to accurate evaluations and treatment plans by detecting soft tissue injuries, joint disorders, and tumors.

Unveiling Infection Examples with CT Sweeps

- To identify diseases in numerous body parts, including the head, chest, belly, and hips, CT scans are necessary because they provide comprehensive cross-sectional images. In the field of disease, CT filters are vital for finding and recognizing cancers. According to the National Lung Screening Trial Research Team (2011), CT scans, for instance, can assist physicians in discovering lung cancer by displaying pulmonary tumors and determining the stage of the disease, allowing them to select the most effective course of treatment.
- CT angiography is extremely helpful for finding conduit blocks and injuries and giving specialists significant data about vascular sicknesses.

Understanding Infection Movement through Visual Data

- The photos that these imaging techniques take are exceptionally useful for figuring out how illnesses deteriorate. Sequential imaging allows specialists to perceive how things change over the long haul, sort out how well medicines are functioning, and make changes on a case-by-case basis.
- Longitudinal MRI studies allow researchers to observe changes in the structure and function of the brain in neurological diseases like Alzheimer's, which provides important information about the progression of the disease.
- In cardiology, ordinary CT angiography assists with following how coronary vein sickness is progressing and assists with arranging therapies to bring down the opportunity of heart issues.

Enhancing Accuracy in Sickness

- Disease control is more accurate thanks to the visual information gathered by these imaging techniques. When specialists make explicit treatment arrangements, they need to know precisely where the sickness is, how awful it is, and what its side effects are.
- In radiation treatment, utilizing both CT check pictures and therapy arranging apparatuses together ensures that growths are unequivocally designated. At the same time, solid tissues around them are harmed as little as could be expected.
- CT angiography helps vascular specialists plan and prescribe correct medicines for conditions like aneurysms or supply route stenosis by allowing them to see the complex vascular organizations.

4. Interdisciplinary Collaboration and Data Integration

Visual data is valuable for cooperating across disciplines and assembling various sorts of information. Multimodal imaging, which incorporates X-beams, X-rays, and CT checks, assists specialists to settle on conclusions about muddled infections. Adding picture information to electronic wellbeing records further develops care progression by allowing specialists to interface visual information with clinical information to get a full image of the patient (Mehmood et al. 2021). By taking a gander at the photos, it can be ascertained how the disorder is deteriorating.

By providing them with useful information about how the body is constantly changing, visual information aids medical professionals in comprehending the progression of diseases. X-rays, computed tomography (CT), magnetic resonance imaging (MRI), and positron emission tomography (PET) are examples of imaging methods that provide clear images of body parts, their functions, and their metabolic processes. Thanks to this, Doctors can now identify issues and observe how a disease progresses over time. Besides the fact that these visual guides assist with finding infections prior, they likewise make it simpler to understand how well medicines function and perceive how far along a sickness is. This data assists specialists with making more customized care designs and works on understanding outcomes. Besides, upgrades in show advances and information examination apparatuses make visual data much more obvious, allowing clinical specialists to get a handle on how illnesses work and settle on shrewd treatment decisions. Table 2 shows the analysis

Table 2. Diseases progression.

Imaging Modality	**Visual Information**	**Contribution to Understanding Disease Progression**	**Examples in Disease Diagnosis**	**Limitations**
X-ray	2D Radiographic images	Provides visualization of bone structures and some soft tissues. Useful for fractures, pneumonia, and lung diseases.	Detecting fractures, pneumonia, lung cancer	Limited soft tissue detail, ionizing radiation exposure
CT (Computed Tomography)	3D Cross-sectional images	Offers detailed views of internal organs and structures. Enhanced resolution for soft tissues.	Staging of cancers, identifying lesions in organs	Ionizing radiation exposure, higher cost compared to X-ray
MRI (Magnetic Resonance Imaging)	High-resolution soft tissue images	Excellent soft tissue contrast without ionizing radiation. Superior imaging of the brain, joints, and organs.	Brain tumors, musculoskeletal disorders, soft tissue injuries	Longer scan times, contraindications for some patients (claustrophobia, metallic implants)
Ultrasound	Real-time imaging using sound waves	Dynamic imaging of internal structures. Safe for real-time monitoring during procedures.	Pregnancy monitoring, abdominal imaging, vascular studies	Operator-dependent, limited penetration for deep structures
PET (Positron Emission Tomography)	Functional metabolic images	Visualizes metabolic activity, detecting changes at a cellular level.	Cancer staging, neurological disorders	Limited anatomical detail, exposure to radioactive tracers
SPECT (Single Photon Emission Computed Tomography)	Functional imaging using radioactive tracers	Evaluates blood flow and organ function.	Cardiac perfusion imaging, bone scans	Lower resolution compared to PET, longer scan times
Mammography	X-ray imaging of breast tissue	Screens for breast cancer, detects microcalcifications and tumors.	Breast cancer screening	Radiation exposure, limited sensitivity for dense breasts

of various imaging modalities for disease progression and the application of these tools for the prognosis of the disease.

5. Image Processing Techniques for Disease Diagnosis

Image processing techniques play a crucial role in disease diagnosis by enabling the extraction, enhancement, and analysis of critical information from medical images. Basic techniques such as noise reduction and contrast enhancement, along with advanced methods like segmentation, feature extraction, and pattern recognition, allow healthcare professionals to detect subtle abnormalities, track disease progression, and derive accurate diagnostic metrics for treatment planning. Integrating AI and ML with image processing further facilitates automated disease detection, classification, and prediction. Overall, these advancements accelerate the diagnostic process and enhance the accuracy of medical image analysis. As innovation continues improving, image-handling strategies could meaningfully alter how sicknesses are analyzed by capitalizing on every one of the advantages of clinical imaging techniques and making it feasible for more customized and successful patient consideration (Thakur and Juneja 2021).

Contribution to disease study: Image processing is significant in handling the data from multiple disease research. It incorporates isolating a clinical picture into clear, significant regions that make it conceivable to find and portray specific designs or issues. The precise division is important for illness identification to find and concentrate on unusual regions in clinical pictures, which makes the accurate area and resulting examination more straightforward.

Methodologies: There are various image segmentation techniques used in disease research. One widely known method is edge-based segmentation, which classifies pixels based on predefined intensity thresholds to detect boundaries. Another approach, region-based segmentation, groups pixels with similar characteristics using methods like region growing, which expands regions from seed points, and region splitting and merging, which divides an image into smaller segments and combines similar regions to enhance accuracy. These techniques play a vital role in isolating key features in medical images, aiding in precise disease detection and analysis. Edge-based segmentation, on the other hand, looks for gaps or changes in levels to find limits. A few further developed techniques are model-based division with dynamic layouts or level sets and gathering strategies like k-implies. Every strategy is intended to work with a particular kind of picture and illness research needs.

5.1 Highlight Extraction Methods and Their Importance in Sickness Forecast

Job in sickness expectation: Highlight extraction is a significant piece of picture handling for sickness expectation since it finds and measures significant data in partitioned regions. The attributes that are separated are taken care of into prescient models, which portrays illnesses and make more precise expectations. Which attributes are utilized straightforwardly influences the discriminative force of prescient models, and that implies it influences how well they can differentiate among great and unfortunate circumstances (Hegde et al. 2022).

Significance in illness expectation: In sickness gauge, different element extraction techniques are utilized to get helpful information from clinical pictures. Measurable attributes like mean, standard deviation, and skewness give us data about how the powers of pixels are fanned out. Texture features, which are made up of how the levels of pixels are arranged in space, can teach you a lot about tissue patterns. Shape-put together elements center with respect to actual properties, while recurrence space highlights, which can be found utilizing techniques, for example, Fourier examination, record highlights of an image in the recurrence range. These techniques are helpful on the grounds that they can transform confounded picture information into mathematical qualities, which makes it simpler to make precise models for anticipating sicknesses (Deng et al. 2022).

6. Classification Methods for Disease Domains

Arrangement techniques are vital in illness regions since they give us coordinated ways of collection things into gatherings and think about what wellbeing related occasions will occur. Computer programs are used in these methods to sort medical data into pre-established groups. This assists with diagnosing illnesses, sorting out their possibilities improving, and arranging medicines.

Application in Sickness Analysis

Clustering techniques are utilized to differentiate between various sicknesses by utilizing characteristics taken from clinical information as information. To place individuals into the right illness bunch, techniques like Help Vector Machines (SVM), Arbitrary Woodlands, and Brain Organizations search for patterns in imaging, clinical, or hereditary information. As a result, treatments that work better and happen faster are made possible through easier early identification and targeted actions (Hammoud et al. 2023).

Prognostic Applications

Clustering techniques can likewise be utilized to think about how a disorder will advance. These strategies use past understanding data to sort out how likely it is that the infection will deteriorate, return, or answer treatment. Prognostic models, which are typically settled on with Choice Trees or Strategic Relapse, assist specialists with making treatment arrangements that are bound to work for every patient, which at last prompts improved results for those patients.

Customized Medication

The development of classification techniques has resulted in personalized medicine, in which medications are tailored to each patient's unique requirements. AI techniques, for example, k-Closest Neighbours and Innocent Bayes, take a gander at qualities that are remarkable to every patient to propose customized medicines. This strategy works on the viability of treatment, brings down the gamble of incidental effects, and utilizes medical care assets (Nguyen et al. 2022).

Improved Symptomatic Precision

Clustering techniques make it a lot simpler to make exact judgments in numerous illness regions. By making use of intricate patterns and connections that can be found in a variety of datasets, these techniques make diagnostic models more sensitive and specific. In order to make estimates that are more accurate and reliable, Ensemble Learning methods like Gradient Boosting and Bagging combine the best features of several models.

Early Recognition and Mediation

At the point when order techniques are utilized, diseases can be seen as right on time, which takes into consideration fast treatment and better persistent outcomes. Utilizing Fake Brain Organizations or Choice Trees to assist with early discovery is significant for beginning medicines from the get-go over an infection, when they are bound to work.

Combination with Multi-Modular Information

Characterization strategies are perfect at assembling information from various sources, similar to clinical pictures, genomes, and clinical records. Using methods like Deep Learning models, you can combine data to get a complete picture of how complicated diseases are. This allows you to make more exact and finish expectations. This brought together strategy works on the exactness of judgments and helps a more extensive arrangement for overseeing patients.

Clinical Choice Emotionally Supportive Networks

Marking frameworks have pursued it feasible for Clinical Choice Emotionally supportive networks (CDSS) to be utilized in medical care. These systems, which make use of algorithms such as Random Forests and Support Vector Machines, aid physicians by providing them with advice that is based on evidence, making it simpler for them to comprehend complex medical data, and ultimately assisting them in making better decisions regarding the care of their patients.

7. Advancements and Challenges in Image-Based Disease Prediction

Picture based sickness forecast has progressed significantly, introducing another period of accuracy medication. New advances are utilized to pull complex examples and signs from clinical pictures so illnesses can be found before they occur. Combining computer vision, DL, and AI has resulted in the development of sophisticated prediction models that can spot minute details in images and connect them to the disease's progression. This improves early diagnosis and prognosis. In any case, these upgrades accompany issues, similar to the way that prescient models should be thoroughly tried and straightforward. There are additionally upright worries about persistent protection and information security, and the calculations could be one-sided or not work the same way for a wide range of patients. The key to personalized, data-driven healthcare models that can enhance patient outcomes and expedite clinical decisions is overcoming these issues and making full use of image-based disease prediction (Wang et al. 2022).

7.1 Recent Innovations in Image Processing for Disease Prediction

As of late, propels in picture handling for illness gauge have made clinical trials more precise and quicker than any time in recent memory. Analysts and specialists can now utilize profound learning, convolutional brain organizations, and move figuring out how to get additional nitty gritty examples and attributes from clinical pictures. This has made it conceivable to make exceptionally precise models that can anticipate numerous sicknesses. Consolidating various sorts of imaging information and adding progressed information improvement strategies has made picture handling calculations surprisingly better at making expectations. This makes it simpler to find, portray, and anticipate complex infections like disease, neurodegenerative issues, and cardiovascular circumstances right off the bat. These novel concepts not only have the potential to improve the accuracy of diagnoses, but they also have the potential to result in personalized, data-driven healthcare treatments that would alter the method by which diseases are predicted and enhance the outcomes of patient care.

Deep Learning Architectures

In the beyond couple of years, picture-based illness expectation has made some amazing progress, with profound learning models driving the way. Convolutional Brain Organizations (CNNs) are truly adept at taking out complex various leveled qualities from clinical pictures, which makes infection figures more exact. Groundbreaking thoughts like consideration cycles and move learning utilize models that have previously been instructed and center around significant pieces of pictures, which prompts better forecast.

Multi-Modal Fusion

A growing trend is to combine data from various types of images to make more accurate predictions about diseases. A disease can be better understood when multiple sources of data, like MRI and PET scans, are combined. These multi-modular strategies, which frequently utilize profound brain organizations, utilize the most amazing aspects of each sense to make a greater assortment for prescient displaying.

Explainable AI in Medical Imaging

Making AI models that can be explained is receiving more attention as the application of AI to medical imaging expands. For specialists to believe programmed sickness expectation frameworks, they should be clear and straightforward. By incorporating attention mechanisms, saliency maps, and other methods that demonstrate how models make decisions, recent enhancements aim to make deep learning models simpler to comprehend.

Quantitative Imaging Biomarkers

Improvements in quantitative imaging biomarkers make it possible to make better predictions about diseases. To get exact information from clinical pictures, AI techniques, for example, relapse models, are utilized. These biomarkers are objective ways of depicting infections and are vital for monitoring how well medicines are functioning.

7.2 Addressing Challenges and Future Directions in this Field

Data Quality and Diversity

One issue that continues to concoct picture-based sickness conjecture is that it needs many top-notch tests. It is difficult to fabricate and utilize forecast models when it is difficult to gain admittance to well-organized data. In the future, it will be necessary to share data, collaborate with other schools, and create standard datasets to ensure that model training is robust and equitable.

Interpretability and Trust

Since profound learning models are 'secret elements', it may be difficult to get specialists to trust them. The objective of future reviews is to make models more obvious by making man-made intelligence strategies that can be made sense of. It is significant for picture-based sickness expectation devices to function admirably with clinical cycles so that specialists can comprehend and trust the outcomes.

Clinical Validation and Integration

Many picture-based sickness expectation models look great, yet they haven't been completely tried and coordinated into clinical practice yet. Later on, huge scope clinical examinations will be utilized to affirm how well these models work, in actuality, medical care circumstances. Making these innovations useful in healthcare necessitates careful consideration of legal issues, as well as their seamless integration into clinical processes.

Ethical Considerations and Bias Mitigation

For fair medical care results, it is critical to manage morals issues and decrease predisposition in picture-based sickness estimate. Guidelines for ethical AI in health should be developed, algorithms should take into account a variety of perspectives, and bias in prediction models should be reduced.

8. Conclusion

The investigation of picture handling in illness estimates and early analysis has uncovered profound experiences and progressive possibilities. Key findings demonstrate the significance of cutting-edge image processing methods for speeding up and improving disease detection. Methods for segmentation, feature extraction, and classification fall under this category. Joining hereditary calculations and AI makes expectations considerably more precise, demonstrating the way that PC strategies and clinical imaging can cooperate. The survey of imaging techniques like X-beams, X-ray, and CT has shown their one-of-a-kind advantage and how they can be utilized in an assortment of sickness regions. Each imaging strategy has its own advantages, for example, X-beams are speedy and modest and X-rays have the option to show extremely itemized pictures of delicate tissues. The mix of various strategies and the utilization of man-made reasoning is a major forward-moving step that will prompt more precise and explicit sickness assessments. Clinical imaging is at the sharp edge of innovation on account of new advancements in picture handling, for example, the ascent of profound learning models, multi-modular combination, and logical artificial intelligence. These upgrades make illness expectations more exact, however they additionally take care of significant issues like trust and intelligibility, making it workable for them to be effectively incorporated into clinical cycles.

As we plan ahead, we want to focus on issues with information quality, clinical confirmation, and morals. People must collaborate to share data, conduct thorough clinical studies, and develop ethical AI models if image processing is to be used responsibly in healthcare. The big picture of how healthcare is changing includes image processing more and more. It is changing illness expectation and early finding into a universe of customized, exact, and patient-focused care. This coming together of PC power and clinical imaging demonstrates the way that vastly improved results for patients can be improved, achieving another period in the field of medication.

References

Arain, Z., Iliodromiti, S., Slabaugh, G., David, A. L. and Chowdhury, T. T. (2023). Machine learning and disease prediction in obstetrics. Current Research in Physiology 100099.

Deng, Y., Qin, H. Y., Zhou, Y. Y., Liu, H. H., Jiang, Y., Liu, J. P. et al. (2022). Artificial intelligence applications in pathological diagnosis of gastric cancer. Heliyon.

Gao, F. (2021). Integrated positron emission tomography/magnetic resonance imaging in clinical diagnosis of Alzheimer's disease. European Journal of Radiology 145: 110017.

Göndöcs, D. and Dörfler, V. (2024). AI in medical diagnosis: AI prediction & human judgment. Artificial Intelligence in Medicine 149: 102769.

Hammoud, M., Kovalenko, E., Somov, A., Bril, E. and Baldycheva, A. (2023). Deep learning framework for neurological diseases diagnosis through near-infrared eye video and time series imaging algorithms. Internet of Things 24: 100914.

He, P., Gao, Y., Shi, L., Li, Y., Jiang, S., Tie, Z. et al. (2023). Motor progression phenotypes in early-stage Parkinson's Disease: a clinical prediction model and the role of glymphatic system imaging biomarkers. Neuroscience Letters 814: 137435.

Hegde, S., Ajila, V., Zhu, W. and Zeng, C. (2022). Artificial intelligence in early diagnosis and prevention of oral cancer. Asia-Pacific Journal of Oncology Nursing 9(12): 100133.

Hu, Z., Wang, Z., Jin, Y. and Hou, W. (2023). VGG-TS winformer: Transformer-based deep learning model for early Alzheimer's disease prediction. Computer Methods and Programs in Biomedicine 229: 107291.

Jiao, C. N., Gao, Y. L., Ge, D. H., Shang, J. and Liu, J. X. (2024). Multi-modal imaging genetics data fusion by deep auto-encoder and self-representation network for Alzheimer's disease diagnosis and biomarkers extraction. Engineering Applications of Artificial Intelligence 130: 107782.

Jomeiri, A., Navin, A. H. and Shamsi, M. (2024). Longitudinal MRI analysis using a hybrid DenseNet-BiLSTM method for Alzheimer's disease prediction. Behav Brain Res. 2024 Apr 12; 463: 114900. doi: 10.1016/j.bbr.2024.114900. Epub 2024 Feb 8. PMID: 38341100.

Jung, S., Ahn, E., Koh, S. B., Lee, S. H. and Hwang, G. S. (2021). Purine metabolite-based machine learning models for risk prediction, prognosis, and diagnosis of coronary artery disease. Biomedicine & Pharmacotherapy 139: 111621.

Kumar, S., Bhagat, V., Sahu, P., Chaube, M. K., Behera, A. K., Guizani, M. et al. (2024). A novel multimodal framework for early diagnosis and classification of COPD based on CT scan images and multivariate pulmonary respiratory diseases. Computer Methods and Programs in Biomedicine 243: 107911.

Mehmood, A., Yang, S., Feng, Z., Wang, M., Ahmad, A. S., Khan, R. et al. (2021). A transfer learning approach for early diagnosis of Alzheimer's disease on MRI images. Neuroscience 460: 43–52.

Nguyen, D., Nguyen, H., Ong, H., Le, H., Ha, H., Duc, N. T. et al. (2022). Ensemble learning using traditional machine learning and deep neural network for diagnosis of Alzheimer's disease. IBRO Neuroscience Reports 13: 255–263.

Prakash, K. and Saradha, S. (2023). Efficient prediction and classification for cirrhosis disease using LBP, GLCM and SVM from MRI images. Materials Today: Proceedings 81: 383–388.

Rahman, M. M., Al-Amin, M. and Hossain, J. (2024). Machine learning models for chronic kidney disease diagnosis and prediction. Biomedical Signal Processing and Control 87: 105368.

Sathiyamoorthi, V., Ilavarasi, A. K., Murugeswari, K., Ahmed, S. T., Devi, B. A. and Kalipindi, M. (2021). A deep convolutional neural network-based computer aided diagnosis system for the prediction of Alzheimer's disease in MRI images. Measurement 171: 108838.

Thakur, N. and Juneja, M. (2021). Early-stage prediction of glaucoma disease to reduce surgical requirements using deep-learning. Materials Today: Proceedings 45: 5660–5664.

Wang, C., Xu, T., Yu, W., Li, T., Han, H., Zhang, M. et al. (2022). Early diagnosis of Alzheimer's disease and mild cognitive impairment based on electroencephalography: From the perspective of event related potentials and deep learning. International Journal of Psychophysiology.

4

Role of Signal Processing in Disease Prediction and Early Diagnosis

1. Introduction

With the advancements in healthcare tools, a new era of personalized and preventative medicine has begun. Signal handling, an expansive region at the convergence of designing and medication, has arisen as a central participant in further developing illness expectation and permitting early revelation. Signal handling does utilize Pcs to take a gander at various types of clinical information. It assists a great deal with finding examples and patterns that are significant for rapidly and accurately diagnosing infections (Vidya and Sasikumaar 2022).

An ever-increasing number of information sources are opening up, similar to clinical pictures, electronic wellbeing records, and information from individual sensors. Our capacity to anticipate diseases may be harmed or helped by this. Signal handling strategies are areas of strength to figure out this information, uncovering secret thoughts that probably would not have been seen previously. With the help of signal processing for disease prediction, early diagnosis and care can be given to patients. Using signal processing for early disease prediction, medical systems can now effectively care for patients (Zhang et al. 2021).

The capacity to further develop testing precision is one of the fundamental motivations behind why sign handling is utilized in illness forecast. Various individuals have various assessments about the consequences of conventional testing techniques, which is the reason the outcomes can be unique. That being said, signal handling strategies take into account unbiased and quantitative examination, which brings down indicative vulnerability and raises the overall precision of diagnostics. This advance toward testing strategies that depend on information could prompt more exact and ordinary infection distinguishing proof.

DOI: 10.1201/9781032722344-4

The focal point of this study goes past the typical areas of clinical analysis and into the area of expectation models. As indicated, signal handling makes it more straightforward to take out complex transient and spatial patterns from longitudinal information. Prediction models that use patient profiles to identify disease risks are now possible because of this. Thus, early assessment turns into a genuine chance, which allows specialists to act rapidly and works on the result for patients (Tang et al. 2022).

Indeed, even with these upgrades, issues should still be fixed before signal handling can be completely utilized to anticipate infections. Information assortment, which comes from various sources and structures, makes it hard to create models that everybody can utilize. To address this issue, progressed signal handling strategies can consolidate and adjust various informational indexes with next to no issues. The occupation is made considerably more earnestly by clinical information containing commotion. As a result, effective signal processing techniques are required to distinguish clinically significant information from noise.

Predicting diseases relies heavily on studying changing amounts of health data in real-time. When signal handling techniques are improved for ongoing handling, they consider consistent following and moment revelation of changes from ordinary wellbeing factors. This expertise is particularly significant for illnesses whose side effects change rapidly since early disclosure can immensely affect how well treatment functions.

Likewise, joining signal handling with man-made brainpower Artificial Intelligence (AI) and Machine Learning (ML) has been extremely useful in tackling troublesome analysis issues. Man-made intelligence driven signal handling models can learn and respond to changing illness patterns all alone, which makes forecast calculations more adaptable and helpful in additional circumstances. The blend of computerized reasoning and information handling techniques is a better approach to concentrate on the most proficient method to foresee illnesses.

In accuracy medication, signal handling helps make clinical medicines more viable by considering the exceptional qualities of every patient. By taking a gander at complex natural signs and information that is remarkable to every patient, signal handling makes it simpler to find individual biomarkers, which thus helps guide centered treatment plans. This customized system is a major change from one-size-fits-all medical services to medical care that is centered around the patient, which is in accordance with the bigger objectives of accuracy medication (Iqbal 2024).

The ethical issues that surface while utilizing signal handling to foresee illnesses are vital. While gathering and examining private wellbeing information, protection issues should be painstakingly and thoroughly examined. Tracking down the right blend between getting valuable data for anticipating sicknesses and safeguarding patients' security is vital and needs progressing study and the making of moral systems.

As sign handling transforms, it is essential to consider what it means for the expense of medical services. Early disease forecasting through signal processing may result in substantial cost savings for healthcare. The monetary viewpoint makes the job of sign-handling illness discovery considerably more significant. This implies that it is not just a logical objective but, in addition, a strategy that can be utilized to set aside cash for medical services frameworks.

Since sickness estimates are so convoluted, engineers, PC researchers, specialists, and medical services specialists must cooperate. Overcoming any barrier between these fields is significant for building total arrangements that influence the information regarding each matter. Cooperating with individuals from various fields ensures that the purposes of sign handling in illness conclusion are by clinical practice and truly help patients (Gupta et al. 2023).

Signal handling is being utilized to foresee illnesses, which is a colossal step in the right direction for medical services. This section looks at the different types of signal-handling techniques. Focusing on how it could change tests, make prescient demonstrating conceivable, and prepare for customized treatment. Before we get excessively profound into the specialized subtleties of utilizing signal handling to anticipate infections, it is critical to know its upsides and downsides. This perspective will assist with melding the fate of medical services through new PC based techniques (Rossini et al. 2020).

1.1 Importance of Early Diagnosis in Disease Management

Early recognizable proof is vital for overseeing sicknesses since it allows individuals to make a move rapidly, which can enormously change how the infection advances. It has been displayed in examinations that early ID further develops treatment results as well as brings down the expense of medical services frameworks. When disease states are discovered quickly, it is possible to use targeted medications, personalized medicine, and more effective treatment methods. This is especially important now that we have precision medicine, in which therapies are tailored to each patient's specific characteristics and are essential for the success of treatments (Jung et al. 2021).

Signal handling is a vital piece of early conclusion since it makes it simpler to track down little examples and patterns in a lot of clinical information. Creative calculations and AI strategies assist with tracking down signs and anomalies that highlight sicknesses in their beginning phases. Early ID makes therapy more compelling, yet it likewise makes it conceivable to do whatever it takes to try not to become ill in any case. Sorting out infections right off the bat allows us to take protection measures, which brings down the opportunity for issues and improves long-haul patient outcomes.

With regard to viral diseases, early discovery is vital for monitoring them. Fast testing apparatuses and early location of breakouts assist general wellbeing authorities with making fruitful strides like quarantine and antibody programs. It is impossible to overstate the significance of early detection for society. Early recognizable proof works on patients' satisfaction by shortening the length and brutality of ailments. It likewise decreases the general interest in medical care assets (Wang et al. 2022).

Signal handling strategies are a helpful device for finding illnesses early, particularly when there are little changes in the body that occur before the side effects appear. These methods, which incorporate seeing clinical pictures and hereditary information, make it conceivable to find indications of illness before they appear.

Early distinguishing proof is particularly significant for ongoing ailments because treating them rapidly can prevent them from deteriorating. This is shown by early medicines for diabetes and coronary illness, which develop glucose control and lower the opportunity of coronary illness.

Early detection is even more important now that smart technology and tools for constant tracking are being used. With ongoing information assortment and investigation, specialists can rapidly detect changes from standard wellbeing factors, which allows them to act rapidly. Getting a determination early is vital for bringing down the expenses of long-haul medicines, emergency clinic stays, and treatment. Early diagnosis helps the economy as a whole because it reduces the impact of diseases on productivity and the ability to work (Zhang et al. 2021).

Advanced wellbeing records and medical care frameworks driven by information have made early location more probable. Assembling patient information from various regions takes into consideration a more intensive examination, which helps find little patterns that probably would not be seen with conventional testing techniques. Identification is significant for something beyond the patient's well-being. Finding normal sicknesses early allows general well-being to advance rapidly, halting the spread of contamination and pandemics.

Signal handling helps track down infections. Additionally, it assists in understanding how well medicines function. Because biomarkers and bodily factors can be continuously examined, treatment can be altered in real time, enhancing its efficacy. Concerning disease, early identification implies better possibilities of life. The Public Malignant Growth Establishment says transmission handling applications like radionics and fluid biopsy examination make disease conclusions more precise. This implies that treatment can begin sooner and better overall view.

2. Fundamentals of Physiological Signals

To precisely distinguish infections, understanding the essential standard of real signs is significant. These incorporate electrocardiography (ECG), electromyography (EMG), and electroencephalography (EEG) (Faraji et al. 2022). These signs are actually similar to how the human body changes after some time. With regards to diagnosing illnesses, ECG readings are vital in light of the fact that they can precisely record heart activity. By dissecting ECG designs, irregularities connected to circulatory illnesses can be found, which significantly helps with the early finding.

Neuromuscular infections can be better perceived with the assistance of EMG readings, which arise from muscle developments. Knowing the subtleties of EMG information helps find issues that could be indications of conditions like myopathies or neuropathies, which makes it simpler to rapidly act. EEG information, which show how the mind is electrically dynamic, are vital for anticipating neurological infections. Breaking down EEG rhythms exhaustively helps find issues associated with epilepsy, mind sicknesses, and psychological wellness issues.

Two important signal-processing techniques that can be utilized to extract useful information from physical data are the Fourier transform and wavelet analysis. These techniques make convoluted signal information more obvious, which is the most vital move toward building precise models for anticipating illnesses. The volume and variety of critical health-related data have increased tenfold since the digitization of medical records and the rise of smart technology. This expansion in their quantity makes it considerably more vital to make progressed signal handling calculations that can deal with a wide variety of data to make precise illness forecasts (Arain et al. 2023).

Signal denoising is a significant stage in the preprocessing system, particularly in physiological sign examination where commotion can conceal little issues. Progressed denoising strategies, such as versatile separating and wavelet thresholding, assist with further developing information quality so that estimates are more precise. Highlight extraction methods, similar to time-space and recurrence space examination, are vital for getting valuable data from substantial signs. These strategies assist with finding remarkable patterns and signs that are significant for illness expectation models.

While AI strategies are joined with signal handling, they have been adept at making expectations about various illnesses. High-level sign handling and AI function admirably together to make precise and adaptable expectation models. Real-time processing of bodily data is essential for rapidly detecting and treating diseases. By integrating parallel processing with edge computing, substantial data can be continuously streamed and analyzed faster and more efficiently. This ensures that you can respond rapidly to new medical conditions.

We have learned a lot about biological systems' complexity from studying nonlinear dynamics in bodily signs. Nonlinear examination strategies, similar to fractal examination and disarray hypothesis, assist with tracking down secret examples in real information. This makes sickness forecast models more precise. A general image of an individual's wellbeing can be acquired from cross-modular examination, consolidating a few real signs. Joining ECG, EMG, and EEG information assists us with completely understanding how the body's capabilities are connected, which makes expectation models more exact (Kamra et al. 2023).

Since utilizing physiological information to foresee illnesses is dishonest, severe protection rules are required. Solid security, safe information keeping, and clear ways of authorization are important to ensure that private wellbeing data is utilized capably. Artificial intelligence and deep learning can be combined for real-time data processing, as indicated by ongoing research. These enhancements may make models more adaptable and exact, stretching the boundaries of how well they can anticipate infections (Al-Maitah 2020).

Table 1 shows a diagram of the essential thoughts behind real signs. It incorporates significant variables like temperature, circulatory strain, breathing rate, and electrocardiogram (ECG). Each signal has a description that explains its physiological significance, how to measure it, its normal range, and any potential clinical effects. In addition, the table depicts the various signals that can be received, how to deal with them, and the challenges that arise when attempting to record and comprehend these bodily factors. This assortment fills in as a fundamental asset for medical services laborers, scholastics, and biomedical specialists, giving a brief yet careful comprehension of the essential standards hidden materially signal following and examination.

Table 1. Fundamentals of physiological signals.

Physiological Signal	**Definition**	**Common Measurement Methods**	**Typical Units**
Electrocardiogram (ECG/EKG)	Records electrical activity of the heart.	Electrodes on the skin, leads connected to an ECG machine.	Millivolts (mV)
Electroencephalogram (EEG)	Measures electrical activity in the brain.	Electrodes attached to the scalp.	Microvolts (μV)
Electromyogram (EMG)	Captures electrical signals from muscles.	Electrodes on the skin above muscles.	Microvolts (μV)
Blood Pressure (BP)	Force of blood against arterial walls.	Sphygmomanometer, automated blood pressure monitor.	Millimeters of Mercury (mmHg)
Respiration Rate (RR)	Number of breaths per minute.	Respiratory belt, nasal thermocouples.	Breaths per minute (bpm)
Skin Conductance (SC/GSR)	Measures sweat gland activity.	Electrodes on the skin.	Microsiemens (μS)
Temperature (Temp)	Body or skin temperature.	Thermocouples, infrared thermometers.	Degrees Celsius (°C)
Pulse Oximetry (SpO2)	Oxygen saturation in the blood.	Pulse oximeter on a fingertip or earlobe.	Percentage (%)
Blood Glucose Level	Concentration of glucose in the blood.	Blood glucose meter.	Milligrams per deciliter (mg/dL)
Electrodermal Activity (EDA)	Sweating response of the skin.	Electrodes measuring skin conductance.	Microsiemens (μS)
Heart Rate (HR)	Number of heart beats per minute.	ECG, pulse oximeter, or photoplethysmography.	Beats per minute (bpm)

2.1 Understanding EEG, ECG, EMG, and Other Physiological Signals

Electroencephalography (EEG) is a vital instrument for anticipating infections since it shows how the cerebrum functions. Analysis has become more precise as a result of recent advancements in the handling of EEG signals, making it possible to identify minute abnormalities indicative of brain diseases (Davies et al. 2023). Figure 1 shows the structure of an Electroencephalography.

Electrocardiography (ECG) signs inform us how the heart functions and are significant for blood wellbeing. Utilizing signal handling on ECG information helps find designs, which makes it simpler to analyze heart issues early and works on the exactness of visualizations (Lucio et al. 2022). Figure 2 shows the signal structure of Electrocardiography.

Electromyography, or EMG, is vital for concentrating on muscle sicknesses. Signal processing techniques are used to process EMG data to identify issues with muscle action. This considers early treatment and better tolerant outcomes (Jayalakshmy et al. 2023). Figure 3 shows the structure of Electromyography.

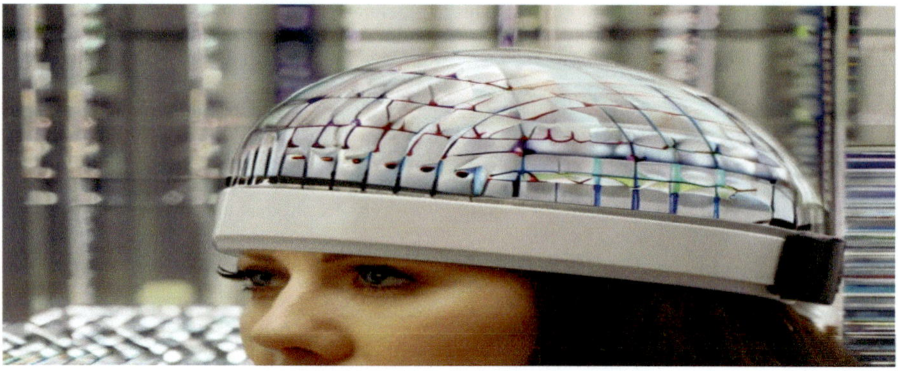

Figure 1. Structure of electroencephalography.

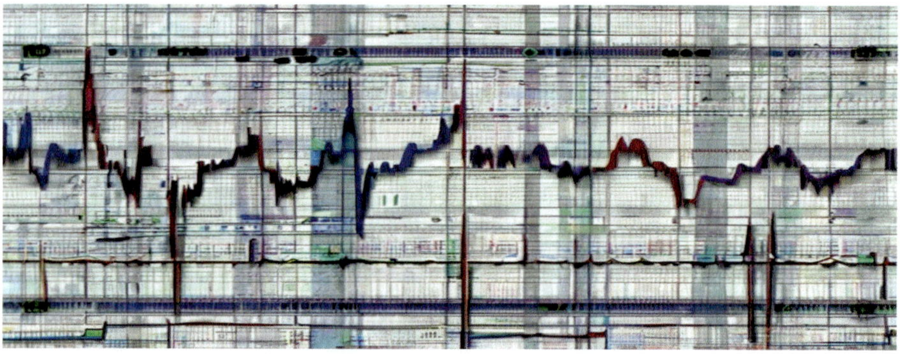

Figure 2. Signal structure of electrocardiography.

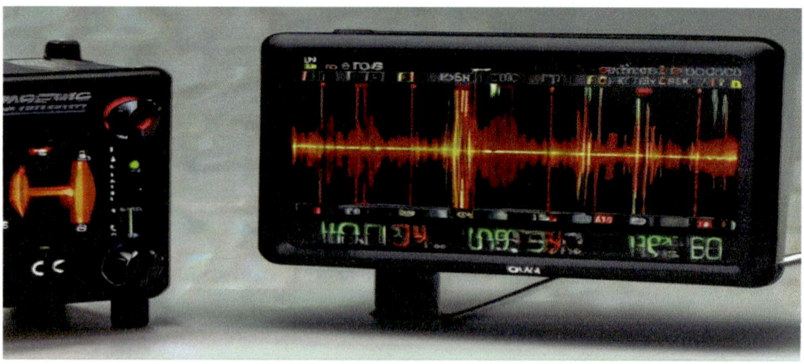

Figure 3. Structure of electromyography.

Assembling EEG, ECG, and EMG information has turned into a more complete method for watching out for wellbeing. Multimodal signal handling procedures assist us with understanding how different physical processes are associated in more detail, which makes it simpler to foresee illnesses. EEG data can be picked up by time-frequency analysis techniques like the wavelet transform and the Hilbert-Huang transform. These methods enable in-depth analysis of EEG data, which aids in the early detection of neural issues.

High level AI procedures are utilized in ECG information handling to track down arrhythmias. Profound learning models, including convolutional brain organizations (CNNs) and repetitive brain organizations (RNNs), have shown uncommon exactness in finding minor ECG designs characteristic of heart irregularities. When EMG data is analyzed using feature extraction techniques, like time-domain and frequency-domain analysis, a lot of insight into how muscles function is gained. This makes it easier to gauge muscle tissues, which assists specialists in finding neuromuscular sicknesses early.

Assembling substantial signs with other well-being-related information, similar to DNA data and way of life factors, makes sickness models better at anticipating what will occur. It is vital to have a full image of an individual's wellbeing while utilizing integrative techniques. In EEG analysis, signal processing systems are increasingly leveraging artificial intelligence (AI) to identify patterns and anomalies. AI-driven models enhance early diagnosis by effectively distinguishing normal brain activity from abnormal patterns.

In ECG signal handling, methods like head part examination (PCA) and recursive element end (RFE) assist with finding the main highlights that can be utilized to distinguish between ECG signals. By zeroing in on significant pieces of the ECG yield, these techniques make determination models work better. Different muscle circumstances cause EMG signs to show specific examples. Techniques for taking care of signs like free part examination (ICA) help to isolate and figure out these examples. This makes demonstrative tests more unambiguous, which prompts more powerful illness-distinguishing proof.

As people develop personal devices for continuous monitoring of physiological signals, they encounter new challenges and opportunities. Real-time data processing on smart platforms enables instant feedback, empowering individuals to take control of their health. It is easier to view a person's entire health history when bodily signal data and electronic health records (EHRs) are combined. Disease models are able to predict the future with greater accuracy by utilizing signal processing on this combined data, allowing for more individualized and targeted treatments (Bron et al. 2021).

We need advanced signal processing techniques to clean and improve the data because bodily signal data is not always accurate due to glitches and noise. A significant piece of working on the sign to-commotion proportion is utilizing versatile separating and apparition expulsion technique to ensure that the symptomatic outcomes are right. Signal handling strategies are continuously changing, and new advances are being added constantly. This will enormously affect how sicknesses are anticipated and viewed right on time. In order to utilize bodily signs to their full potential in enhancing personalized healthcare, researchers and practitioners must keep up with these changes.

2.2 Significance in Disease Prediction and Diagnosis

It is vital in the present medical services framework to utilize signal handling to foresee illnesses and make early analyses. It is essential for enhancing diagnostic abilities. By combining various types of data sources, such as medical images, genetic information, and clinical records, signal processing makes it simpler to locate the intricate patterns necessary for accurate disease recognition.

The improvement of PC based strategies in signal handling has raised the field higher than ever, making it conceivable to examine immense datasets at speeds and with more exactness than ever before. This PC expertise makes indicative models more exact at foreseeing what will occur, prompting a push toward deterrent and customized medical care procedures. Something significant that sign handling does is select little biomarkers that show beginning phase sicknesses, so therapy can start on time. High level calculations can track down tiny changes in substantial signs. This could give us a method for finding infections from the get-go.

Notwithstanding standard testing strategies, new advances like shrewd tech and Web of Things (IoT) screens make signal handling vital. These devices continue to convey wellbeing related information constantly, so mind boggling signal handling techniques are expected to figure out everything and work on the precision of infection forecasts. It is vital to utilize signal handling to find irresistible infections right on time by taking a gander at the study of disease transmission information and patterns. Ongoing handling of various sorts of information, for example, information from virtual entertainment and portability applications, allows us rapidly to detect potential assaults and find proactive ways to safeguard general wellbeing.

A major piece of accuracy medication is utilizing signal handling to sort individuals into bunches in light of their sub-atomic and hereditary characteristics. This arrangement works on the precision of illness expectations, which allows specialists to make more customized treatment designs that have less side effects. Neurodegenerative illnesses make signal handling considerably more significant in light of the fact that they are difficult to analyze early. When neuroimaging information is handled by cutting edge signal handling strategies, little changes in cerebrum construction and work can be found. This gives researchers significant data about illnesses like Alzheimer's (Siddiqui et al. 2023).

By creating systems that are capable of learning on their own, artificial intelligence (AI) and signal processing enhance their utility for disease prediction. The prescient abilities of these artificial intelligence improved models are continuously growing, responding to how infections change over the long run and making analyses more precise. Signal handling is significant for managing various kinds of information since it makes expectation models more solid across many patient gatherings. Predictive models can be utilized in real-world healthcare settings thanks to the fact that adaptive signal processing techniques are able to deal with shifts in the distribution of data.

The appearance of 5G innovation makes signal handling significantly more important in anticipating and diagnosing sicknesses. Quick information moves and lower postpone make it conceivable to dissect wellbeing information progressively, which assists individuals to opt for speedy choices in basic medical care circumstances. Signal handling added to telemedicine stages analyzes issues from a remote place, separating actual limits and making medical care benefits more open. Remote observing of physiological signals and high-level sign handling let specialists come to conclusions about their patients without actually being near them.

Multimodal information combination, in which information from various sources is consolidated for a more complete demonstrative technique is likewise a significant piece of sign handling. Assembling signs from pictures, hereditary qualities, and clinical information works on our general comprehension of sickness states and makes forecast models more precise. Signal handling calculations assist with making prescient signs that assist specialists with finding individuals who are bound to get specific infections. These biomarkers serve as early warning signals, enabling specialists to conduct targeted screenings and take proactive measures to prevent potential health issues.

Since illnesses change over time, they should be observed constantly. For ongoing disease tracking, signal processing excels at handling time-series data. These frameworks assist with the versatile therapy of persistent circumstances by getting little changes over the long haul.

3. Techniques for Signal Analysis in Disease Prediction

Signal examination is a vital piece of foreseeing illnesses since it utilizes a great many techniques to get helpful data from various substantial signs. This study investigates three significant areas of sign examination used to anticipate illnesses: signal-sifting techniques, highlight extraction strategies, and characterization techniques. Signal separating is significant for working on the sign-to-commotion proportion in real information, which is expected to find sicknesses precisely. Utilizing strategies like bandpass and versatile sifting decreases disarray, making it more straightforward to choose illness-explicit examples from the commotion.

Productively separating highlights is the main move toward transforming crude real information into valuable models for anticipating sicknesses. Time-area and recurrence space characteristics, similar to adequacy, recurrence, and fluctuation, give us valuable data about the properties of substantial signals that we can use for additional review. In order to transform the collected features into useful insights for disease prediction, the application of classification methods is essential. Support vector machines, brain organizations, and choice trees are a portion of the AI strategies that are vital for arranging signals into various illness states in view of the elements that were recuperated.

Methods for signal filtering, feature extraction, and classification, all work well together to create a procedure that enables the creation of robust disease forecast models. This consolidated strategy works on the accuracy and steadfastness of infection expectation models by cautiously eliminating clamor, taking out significant highlights, and utilizing progressed arrangement calculations. The use of signal analysis to predict diseases has made progress, but there are still issues, like the need for real-time processing and the need to adjust to changes in the body. New algorithms and mixed methods that combine AI and signal analysis are still being investigated by researchers in order to address these issues and enhance prediction accuracy.

With the abundance of smart technology devices available, signal analysis now enables real-time monitoring of physiological signs for continuous health assessment. Signal handling procedures in wearables permit the extraction of significant highlights, making room for on-the-fly sickness conjecture and customized wellbeing therapies. Accuracy medication focuses on the requirement for customized signal examination, and that implies making sickness expectation models fit the special examples of every patient. Tweaked highlight extraction and arrangement strategies make it conceivable to recognize and determine sicknesses all the more precisely and to have the patient as a main priority.

Worries about morals become more significant as sign examination is involved more in medical care choices. Analysts and professionals should ensure that protection, informed assent, and clear correspondence are completely regarded while involving individual wellbeing information in signal examination models.

Problem: To sort out when a specific circulatory sickness will begin, a review group is checking out substantial signs from a gathering of patients. The group has accumulated electrocardiogram (ECG) information from individuals who are sound and individuals who have been marked with the sickness. They want to find unique characteristics in the ECG data using signal processing techniques in order to accurately predict diseases (Azani et al. 2023).

Given Information

- The ECG readings are saved as discrete time series data, which are shown as X(t) for healthy people and Y(t) for people who have been identified with the disease. T seconds is how long the ECG signs last.
- The team's goal is to get useful information from the signals so they can tell the difference between healthy and sick states.
    - **Signal Filtering:** Apply a bandpass filter to both X(t) and Y(t) signals to enhance the frequency components relevant to cardiovascular abnormalities. Represent the filtered signals as $X_f(t)$ and $Y_f(t)$.
    - **Feature Extraction:** Calculate time-domain features from the filtered signals, such as mean, standard deviation, and skewness. Represent these features for the healthy group as μ_x, σ_x, and $Skew_x$ and for the diseased group as μ_y, σ_y, and $Skew_y$.
    - **Classification:** Use a support vector machine (SVM) or another classification method to tell the difference between healthy and sick states based on the traits that were retrieved. Use the right measure to figure out how accurate the classification model is.

Solution

- *Signal Filtering*: Utilize a bandpass filter with a passband between f_{low} and f_{high} Hz. The filtered signals are obtained using the convolution operation:

$$X_f(t) = X(t) * h(t),\ Y_f(t) = Y(t) * h(t)$$

- Feature Extraction: Calculate the time-domain features as follows:

$$\mu_x = \frac{1}{t}\int_0^T X_f(t)dt, \sigma_x = \sqrt{\frac{1}{T}\int_0^T (X_f(t) - \mu_x)^2 dt},$$

$$Skew_x = \frac{1}{T}\int_0^T \left(\frac{X_f(t) - \mu_x}{\sigma_x}\right)^3 dt$$

  Similarly, calculate μ_y, σ_y, and $Skew_x$ for the disease group.
- Classification: Train an SVM classifier using the extracted features and evaluate its accuracy on a separate validation dataset.

Figure 4 shows how cutting-edge signal-explicit calculations are utilized to catch and deal with physiological information. It additionally indicates the best way to recognize, include, and imagine physiological signs. Eventually, these calculations catch exceptional physiological elements by doing many errands, for example, quality checks, commotion separating, phantom evacuation, division, and trademark point acknowledgment on biosignal waveforms. At the point when you can see these highlights, you can gain tons of useful knowledge about the muddled examples and qualities of physiological signs. Understanding and interpreting these signals is crucial for comprehending the physiological events they represent. As a component of the element extraction process, genuinely huge minutes and time-recurrence examinations are determined. The objective is to isolate sincerely significant highlights from physiological signs like electroencephalogram (EEG),

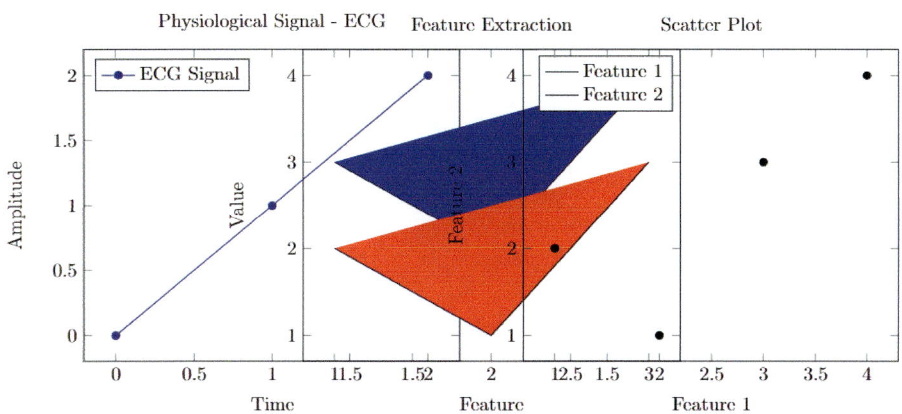

Figure 4. Feature extraction and physiological signal visualization.

galvanic skin reaction (GSR), electromyogram (EMG), skin temperature, breath example, and blood volume beat. This will assist us with diving deeper into how these signs collaborate with one another.

4. Intelligent Systems Integration

The utilization of smart frameworks, particularly man-made reasoning (Computer-Based Intelligence) and AI (ML) techniques, has impacted how information is dissected and flags are handled for sickness estimates. This study investigates two vital parts of insightful framework coordination: The application of advanced techniques in data analysis, along with the pivotal role of AI and machine learning in signal processing, enhances disease prediction.

Current information investigation for sickness figure incorporates shrewd techniques like information mining, design acknowledgment, and information portrayal as significant parts. These techniques assist with getting helpful data from extremely huge and confounded medical care datasets, tracking down examples and associations that ordinary examination strategies could miss.

Computer based intelligence and ML have become exceptionally helpful in signal handling. They have changed how sicknesses are anticipated by having the option to adjust, learn, and track down complex examples in real information. AI calculations like profound learning and outfit strategies consequently find significant qualities and associations in information, which makes illness models more precise at foreseeing what is in store.

Figure 5 shows how man-made reasoning (simulated intelligence) and AI (ML) have totally altered how sicknesses are anticipated. These new innovations have fundamentally altered how medical services are finished by making it conceivable to make exceptionally precise expectation models. Overwhelming patient information, including hereditary data, clinical pictures, electronic wellbeing records, and way of

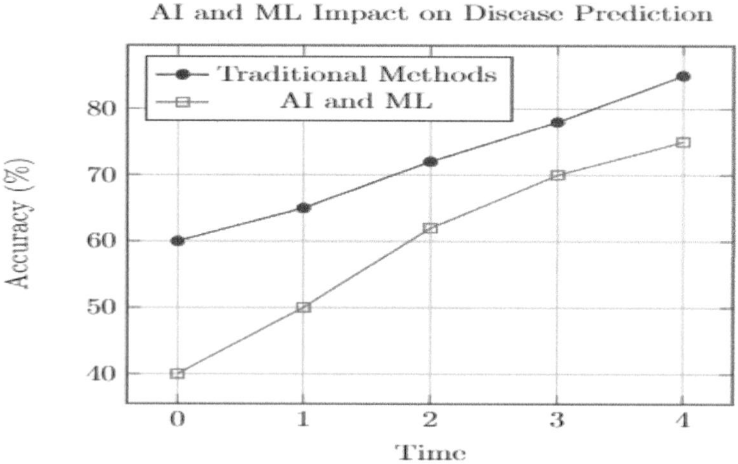

Figure 5. Comparison of accuracy over time.

life variables, computer-based intelligence and ML frameworks can find complex examples and signs that frequently get away from conventional analysis techniques. Also, these advancements permit specialists to think about when various sicknesses will begin, how they will advance, and what could occur eventually. This makes early inclusion and customized treatment plans more straightforward. Be that as it may, utilizing computer-based intelligence and ML to distinguish illnesses accompanies issues like information security, calculation predisposition, and having the option to grasp the outcomes. These issues should be painstakingly and thoroughly examined and dealt with in a mindful manner to ensure that estimate models are dependable and incorporate everybody. In any case, artificial intelligence and ML's extreme capacity in sickness expectation holds expect working on tolerant outcomes, utilizing assets, and pushing medical services toward a more proactive and customized approach.

Intelligent frameworks join various kinds of information sources, similar to clinical pictures, clinical records, hereditary qualities, and shrewd gadget information, so they can give a full image of an individual's wellbeing. This overall strategy makes it simpler for sickness models to make expectations, which prompts more precise and explicit gauges. Adding information demonstrating strategies like ontologies and semantic models to savvy frameworks makes infection forecast models more obvious. By organizing clinical data systematically, these devices make it easier to understand decision-making processes and provide predictions with greater contextual relevance.

Intelligent systems enable real-time decision support, allowing doctors and nurses to make quick, well-informed decisions based on continuously evolving patient data. This adaptable strategy is particularly significant for anticipating sicknesses, where fast activities can hugely affect results. Building trust and acceptance in the healthcare community requires ensuring that intelligent systems used for disease forecasting can be explained to everyone and are readily accessible. Clinicians can comprehend the reason why forecasts are made when man-made intelligence models are made that can be deciphered. This assists individuals with cooperating to make decisions.

Despite the fact that smart frameworks can make a huge difference, they are still difficult to use for sickness expectation due to social worries, information security issues, and the requirement for uniform methodology. Getting these issues tackled is essential for clever frameworks to be generally utilized in clinical practice. Future disease identification will require the development of mixed intelligent systems that effectively integrate domain-specific information, AI, and machine learning (Kaswan et al. 2022). This consolidating of various strategies makes prescient models more solid by utilizing the most awesome aspects of each to get around their imperfections.

To effectively coordinate smart frameworks, medical care laborers should cooperate to ensure that they are in accordance with clinical cycles and that subject information is utilized. This joint technique works on the genuine helpfulness and acknowledgment of sharp frameworks in sickness recognition.

5. Application of Signal Processing in Disease Prediction

Using signal processing for disease prediction is a groundbreaking approach that employs advanced techniques to extract valuable insights from diverse data sources. This concentrates mostly on contextual investigations and how they can treat various infections. It shows both the victories and issues that surface in actuality. By concentrating on electrocardiogram (ECG) information, signal handling has shown astounding outcomes in foreseeing heart diseases. Contextual analyses show that it is feasible to find little issues in heartbeats, which permit early administration and prevent perilous occasions from occurring.

In the realm of disease finding, signal-handling techniques applied to clinical imaging information show confident outcomes. A few triumphs incorporate finding disease right on time by seeing little subtleties in x-beams, which opens up better approaches to redo treatment plans. Signal processing is used to predict brain diseases by analyzing electroencephalogram (EEG) data for conditions like seizures. Examples of overcoming adversity show how accurately recognizing pre-seizure examples can prompt early treatment measures and improve patient results.

To utilize signal handling to anticipate irresistible infections, various kinds of information should be investigated, such as web-based entertainment patterns and clinical information. In reality, applications show that it is feasible to anticipate illness flare-ups, which assists with general wellbeing responses and asset sharing. As signal handling techniques are utilized on constant glucose, the following information has shown a guarantee of recognizing diabetes-related issues. Personalized control strategies are made possible by discovering case studies of glycemic trends that indicate potential issues.

Signal processing on breathing signs and pulmonary function tests can assist in identifying lung diseases like COPD. One example of overcoming adversity is the early recognition of demolishing lung capability, which is considered a proactive measure to lessen eruptions. Despite the fact that there have been triumphs, signal handling for sickness recognition is still difficult to use in reality. Boundless utilization is difficult to accomplish as a result of issues like information sharing, principles, and the requirement for solid assessment frameworks. It is vital to take care of these issues so that sign handling can be handily utilized in clinical settings.

It is vital to interface signal handling to electronic wellbeing records (EHR) with the goal that expectation models can be utilized for a huge scope and for quite a while. Effective applications demonstrate the way that signal-determined experiences can be effortlessly joined with full tolerant wellbeing information, which makes gauges more helpful in reality. A vital piece of genuine execution is ensuring that infection gauges are exact for every patient. Finding the right blend between individual forecasts and generalizability across many patient gatherings is hard, which is the reason why study and model improvement are progressing.

6. Future Directions and Challenges

Signal handling is still vital for anticipating infections, so investigating the future and anticipating issues is vital for pushing this region ahead. This discussion is for the most part about recent fads and the moral and specialized issues that should be addressed before signal handling can be utilized to anticipate illnesses.

Later on, there will presumably be significantly more converging of multi-modular information in signal handling for sickness location. Hereditary qualities, pictures, savvy tech, and electronic wellbeing information are only a portion of the wellsprings of data that could be utilized to get a superior image of illness states. In this turn of events, it will be vital to have progressed signal handling techniques that can deal with and get valuable data from these different datasets.

As models for man-made brainpower (computer-based intelligence) and AI (ML) get more confounded, they should be straightforward. As indicated, the future will see the production of sign-handling models that accentuate being clear and straightforward. Healthcare professionals will be able to comprehend better and accept the decisions made by these systems due to this. Utilizing simulated intelligence that can be made sense of ensures that results are right and straightforward in a clinical setting.

The eventual fate of sign handling in sickness estimates incorporates the capacity to ceaselessly watch an individual's well-being and give continuous data about their condition. This shift toward continuous, dynamic analysis enables the rapid detection of subtle changes, allowing immediate intervention and personalized adaptation to each individual's needs. These advancements can potentially transform disease management, particularly for chronic conditions.

Edge registering may be more significant for signal handling than thinking about information protection, security, and the requirement for continuous handling. It is feasible to handle private well-being information without sending huge records to focal PCs by breaking down information locally on gadgets or at the organization's edge. Both privacy and speed, which are crucial in healthcare apps, are improved by this method.

Due to ethical and technological issues, using signal processing to predict diseases is still difficult. A major issue is ensuring information quality and consistency across various clinical circumstances. Various information assortment strategies, structures, and quality make it hard to make viable forecast models that everybody can utilize.

Considering social issues like security stresses and potential defects in forecast models is vital. Pushing ahead, we ought to zero in on areas of strength for building that can decrease predispositions and ensure that sign-handling models are fair, particularly when they are utilized for underserved and shifted gatherings.

A continuous issue is making it feasible for various medical services frameworks and devices to cooperate with no issues. The future of signal processing for disease prediction requires open interfaces, standardized data forms, and protocols. This will permit prescient models to be utilized in various medical service settings. Ensuring that legitimate principles are followed and getting affirmations become vital as sign handling models become normal in clinical practice. Making uniform principles and administrative frameworks will be critical to sort out the security, adequacy, and moral issues associated with utilizing signal handling models in medical services settings.

Later on, individuals and machines will cooperate such that they settle on signal-handling models more like choice help apparatuses than discrete elements. When medical care laborers and brilliant frameworks cooperate, they can utilize the most amazing aspects of one another to make illness forecast models more exact and valuable. Getting individuals to be aware of and acknowledge the utilization of sign handling in illness discovery is vital for making it work. Teaching people in general about the advantages, constraints, and social contemplations encompassing forecast models will be significant in getting trust and guaranteeing wide acknowledgment in medical services rehearses.

7. Conclusion

The investigation of sign handling in sickness finding has prompted numerous significant disclosures and significant advances in the field of medical care. The use of cutting-edge signal processing techniques has made it much easier to spot minute patterns in a variety of physical signs, enabling prompt and precise diagnosis of a wide range of illnesses. The success stories of case studies, ranging from heart problems to infectious diseases, demonstrate how signal processing could alter the delivery of healthcare. The coming together of various sorts of information, the improvement of computer-based intelligence that can be made sense of, and the continuous following of wellbeing signals are viewed as three significant patterns that will shape the fate of sign handling in illness expectation. With the chance of ongoing activities and the expansion of edge PC devices, this strategy has significantly more potential to change things. In general, these outcomes show that medical services are moving toward exactness, proactive consideration, and customized care. Signal handling could meaningfully affect infection gauge that goes past how well it can analyze the sickness. It has the potential to alter the delivery of healthcare by facilitating early measures, minimizing adverse effects of treatment, and enhancing patient outcomes. Combining smart systems, ethical concerns, and human-machine cooperation may be the key to overcoming operational obstacles and ensuring compliance with ethical and legal standards for these novel concepts.

There are a great number of fascinating things that could occur in the future with signal handling and sickness gauges. Signal handling techniques are continuously changing, and innovation is improving. This will change the field of prescient medication as we push toward a period of customized and information-driven medical services. Effectively utilizing these innovative concepts while upholding ethical considerations is crucial to maximizing the benefits of signal processing and ensuring a positive global impact on healthcare.

References

Al-Maitah, M. (2020). RETRACTED ARTICLE: Analyzing genetic diseases using multimedia processing techniques associative decision tree-based learning and Hopfield dynamic neural networks from medical images. Neural Computing and Applications 32(3): 791–803.

Arain, Z., Iliodromiti, S., Slabaugh, G., David, A. L. and Chowdhury, T. T. (2023). Machine learning and disease prediction in obstetrics. Current Research in Physiology 100099.

Azani, A., Omran, S. P., Ghasrsaz, H., Idani, A., kadkhodaei Eliaderani, M., Peirovi, N. et al. (2023). MicroRNAs as biomarkers for early diagnosis, targeting and prognosis of prostate cancer. Pathology-Research and Practice 154618.

Bron, E. E., Klein, S., Papma, J. M., Jiskoot, L. C., Venkatraghavan, V., Linders, J. et al. (2021). Cross-cohort generalizability of deep and conventional machine learning for MRI-based diagnosis and prediction of Alzheimer's disease. NeuroImage: Clinical 31: 102712.

Davies, M. P., Sato, T., Ashoor, H., Hou, L., Liloglou, T., Yang, R. et al. (2023). Plasma protein biomarkers for early prediction of lung cancer. eBioMedicine 93: 104686.

Faraji, G., Moeini, P. and Ranjbar, M. H. (2022). Exosomal microRNAs in breast cancer and their potential in diagnosis, prognosis and treatment prediction. Pathology-Research and Practice 154081.

Gupta, R., Kumari, S., Senapati, A., Ambasta, R. K. and Kumar, P. (2023). New era of artificial intelligence and machine learning-based detection, diagnosis, and therapeutics in Parkinson's disease. Ageing Research Reviews 102013.

Iqbal, J. M. (2024). A novel deep learning approach for early detection of cardiovascular diseases from ECG signals. Medical Engineering & Physics 125: 104111.

Jayalakshmy, S., Lakshmipriya, B. and Sudha, G. F. (2023). Bayesian optimized GoogLeNet based respiratory signal prediction model from empirically decomposed gammatone visualization. Biomedical Signal Processing and Control 86: 105239.

Jung, S., Ahn, E., Koh, S. B., Lee, S. H. and Hwang, G. S. (2021). Purine metabolite-based machine learning models for risk prediction, prognosis, and diagnosis of coronary artery disease. Biomedicine & Pharmacotherapy 139: 111621.

Kamra, V., Kumar, P. and Mohammadian, M. (2023). An intelligent disease prediction system for psychological diseases by implementing hybrid hopfield recurrent neural network approach. Intelligent Systems with Applications 18: 200208.

Kaswan, K. S., Baliyan, A., Dhatterwal, J. S., Jain, V. and Chatterjee, J. M. (2022). Intelligent classification of ECG signals using machine learning techniques. Healthcare Monitoring and Data Analysis Using IoT: Technologies and Applications, Decision and Control 412: 1–16.

Lucio-Gutiérrez, J. R., Cordero-Pérez, P., Farías-Navarro, I. C., Tijerina-Marquez, R., Sánchez-Martínez, C., Ávila-Velázquez, J. L. et al. (2022). Using nuclear magnetic resonance urine metabolomics to develop a prediction model of early stages of renal disease in subjects with type 2 diabetes. Journal of Pharmaceutical and Biomedical Analysis 219: 114885.

Rossini, P. M., Di Iorio, R., Vecchio, F., Anfossi, M., Babiloni, C., Bozzali, M. et al. (2020). Early diagnosis of Alzheimer's disease: the role of biomarkers including advanced EEG signal analysis. Report from the IFCN-sponsored panel of experts. Clinical Neurophysiology 131(6): 1287–1310.

Siddiqui, S. A., Ahmad, A. and Fatima, N. (2023). IoT-based disease prediction using machine learning. Computers and Electrical Engineering 108: 108675.

Tang, S., Yuan, K. and Chen, L. (2022). Molecular biomarkers, network biomarkers, and dynamic network biomarkers for diagnosis and prediction of rare diseases. Fundamental Research.

Vidya, B. and Sasikumar, P. (2022). Parkinson's disease diagnosis and stage prediction based on gait signal analysis using EMD and CNN–LSTM network. Engineering Applications of Artificial Intelligence 114: 105099.

Wang, C., Xu, T., Yu, W., Li, T., Han, H., Zhang, M. et al. (2022). Early diagnosis of Alzheimer's disease and mild cognitive impairment based on electroencephalography: From the perspective of event related potentials and deep learning. International Journal of Psychophysiology.

Zhang, M., Cheng, S., Jin, Y., Zhao, Y. and Wang, Y. (2021). Roles of CA125 in diagnosis, prediction, and oncogenesis of ovarian cancer. Biochimica et Biophysica Acta (BBA)-Reviews on Cancer 1875(2): 188503.

Zhang, Y., Wang, Y., Zhang, B., Li, P. and Zhao, Y. (2023). Methods and biomarkers for early detection, prediction, and diagnosis of colorectal cancer. Biomedicine & Pharmacotherapy 163: 114786.

5

Impact of Behavioral Analysis in Prediction

1. Introduction to Behavioral Analysis in Disease Prediction

New approaches to using behavioral analysis to identify and comprehend how diseases change over time have been made possible by the advancement of sophisticated computer methods. Combining machine learning algorithms and big data analytics has made it much simpler to decipher intricate behavioral patterns and predict future health outcomes. Conduct examination is utilized to foresee illnesses since the activities and lifestyle decisions of individuals immensely affect their well-being. By utilizing PC strategies, scientists can investigate documents that contain an individual's advanced records. This provides them with an image of the individual's day to day exercises, social contacts, and exposures to climate (Huang et al. 2019).

To completely comprehend how conduct and sickness are associated, a gander at numerous perspectives, like living variables, social drivers, and natural impacts needs to be taken. Joining various kinds of information, such as electronic wellbeing records, and virtual entertainment conduct together with utilizing computational models, makes it simpler to get a full image of the numerous factors that influence the risk of sickness. Profound brain organizations and outfit strategies are two instances of AI procedures that can be utilized to track down little patterns in extremely enormous arrangements of social information. These calculations are truly adept at finding joints that don't follow a straight line. They are a dynamic framework for disease prediction models because they can change in response to shifting behavioral trends.

Conduct examination goes past the standard study of disease transmission strategies since it gives ongoing data about how populaces change and what dangers individuals face. By continually gathering and dissecting social information, AI models identify hidden patterns. This enables quick identification of new health risks and protects individuals from health issues. Adding spatial examination to sickness expectation models makes them significantly more exact because it considers

DOI: 10.1201/9781032722344-5

contrasts in conduct and wellbeing results in the particular area. The utilization of Geographic Data Frameworks (GIS) makes it conceivable to plan areas of interest of illnesses, which assists with centered general wellbeing measures and asset assignment in light of how individuals act in specific regions (Shalini 2021).

With regards to security and information assurance, conduct examination in illness estimates brings up moral issues. Tracking down the right blend between getting helpful data and safeguarding individuals' security is significant for ensuring that conduct information is utilized in a dependable and fair manner for prescient displaying. Wearable tech and the Web of Things (IoT) have become extremely helpful to get constant data about how individuals act. These gadgets continually track substantial elements, practice levels, and rest propensities. This adds to conduct datasets and makes prescient models more precise (Alahmadi and Soh 2011).

Conduct phenotyping is another area of PC wellbeing concentrate that takes a gander at how individuals can be placed into bunches, in view of how they act and respond to various factors. This strategy makes it conceivable to get a more exact and customized image of how individuals are probably going to get sicknesses. It also makes it feasible for customized medicines and accuracy medication intends to be made. Prescient models that utilize conduct investigation have been truly adept at anticipating instances of irresistible illnesses. By taking a gander at patterns of how individuals associate with one another, how they move, and how they share data, PC models can offer early hints and assist general wellbeing by finding a proactive way to stop the spread of dangerous illnesses.

Adding mental variables to models that foresee illness makes them more confounded and nuanced. This is on the grounds that psychological wellness and actual wellbeing are connected in the two bearings. We can learn more about the factors that influence the progression of diseases by utilizing computer methods that take psychological variables into account. Conduct investigation isn't simply used to make expectations about people; it can likewise be utilized to make expectations about the wellbeing of entire networks through informal organization examination. Computational models can track down significant hubs and utilize social elements to unveil wellbeing intercessions more powerfully by taking a gander at the design of social associations and data stream inside networks (Dou et al. 2024).

Despite the fact that PC techniques have improved, it is still difficult to join various kinds of information and make illness figure models that can be utilized by everybody. To utilize conduct examination to its fullest potential in foreseeing sicknesses on a worldwide scale, we really need to cooperate on normalization and joint tasks.

1.1 The Role of Human Behavior in Disease Prediction

The part that the conduct of individuals plays in foreseeing sicknesses is a perplexing region that joins clinical information with patterns in conduct. Late examinations show that living decisions, accepting prescriptions as recommended, and social communications hugely affect wellbeing results. For expectation models in medical services, knowing the significance of social exploration is developing constantly.

Social information, which incorporates elements like activity, food, and feelings of anxiety, gives a more intricate picture that accompanies standard clinical information. This makes infection expectation frameworks more exact (Abdeen et al. 2021).

When used to recognize sicknesses, social investigation checks out something beyond a singular's activities. It additionally examines the social factors that influence health. It is helpful to examine how people are affected by social networks, financial factors, and neighborhood effects in order to get a complete picture of how health behaviors change over time. Utilizing conduct examination in medical care frameworks makes it simpler to find early advance notice signs, which allows individuals to make a move and stop issues before they deteriorate. AI frameworks utilize conduct information to foresee who will get an infection and how quickly it will spread.

While taking insights from human conduct and behavioral patterns for foreseeing infections, it tends to be utilized as an adaptable and changing contribution for models that make expectations. Customized wellbeing gauges are made conceivable by continuously following individuals conduct with savvy tech and portable applications. In medical care, conducting an examination does not just predict an individual's chance of getting a sickness, but it is also a vital piece of dealing with the well-being of an entire local area. Designated activities and general well-being endeavors can be made conceivable by tracking down comparative examples of conduct among specific gatherings (Rane 2018).

Prescient demonstrating has moved into another level with the ascent of huge information examination in medical services. By adding conduct information to huge wellbeing datasets, gauges have become more precise, prompting a superior comprehension of how sicknesses progress. Models that attempt to predict diseases rely heavily on psychological factors in behavioral research. Decision-making, perception, and mental health status are all examples of cognitive factors that can reveal a lot about a person's likelihood of developing particular health issues.

The fact that behavioral analysis can uncover hidden patterns and connections in healthcare data is crucial. It is possible to extract useful information from large sets of behavioral data by combining advanced statistical methods with machine learning techniques. Social examination empowers a patient-focused way to deal with infection expectation, which considers that every individual's wellbeing process is unique. This individualized view is vital for ensuring that counteraction and treatment plans are custom fitted to the one-of-a-kind requirement of a great many patient gatherings.

Since conduct information is so point by point, it is feasible to find little changes that occur before illnesses appear in the body. An early disclosure through social examination makes deterrent estimates bound to work, which could steer ailments. The best utilization of medical services assets is helped by prescient models in light of social examination. By sorting out which individuals and gatherings are at the most elevated risk, medical care laborers can utilize their assets to ensure that the correct individuals get help when they need it. This will bring down the absolute expense of medical services (Geeitha et al. 2023).

By interfacing clinical and non-clinical information, conduct investigation assists us with getting a full image of the elements that influence wellbeing. Adding various kinds of information to sickness expectation models makes them more precise and dependable, which prompts better medical services choices. Conduct research is valuable for anticipating infections, all things considered, including long haul ones. Conduct can be followed after some time to track down drifts that lead to the development and spread of constant infections. Doctors are able to tailor treatments to enhance control as a result.

Adding social examination to models that foresee illnesses is connected with the possibility of accuracy medication. Medical services arrangements can be made to fit the particular necessities of every patient by considering their propensities and interests in the most effective way to develop therapy results further. Social examination sees how individuals' activities change over the long haul, which makes it a crucial method for assessing risk. A longitudinal investigation of social information works on prescient models over the long haul by uncovering risk factors that change after some time.

At the point when conduct examination is utilized to foresee infections, moral issues must be pondered upon. Strong ethics models and rules are needed to strike a balance between the benefits of better prediction and privacy concerns. The use of behavioral analysis to predict diseases is in line with the trend toward proactive and preventative healthcare strategies. Through conduct experiences, medical services frameworks can move from responding to and lessening wellbeing gambles, becoming proactive in their consideration.

To take advantage of conduct research for infection location, individuals from various fields need to cooperate. Data scientists, healthcare professionals, and behavioral scientists need to work together to create models that fully reflect the complexity of human behavior. The utilization of human conduct in sickness expectation devices is a major forward-moving step in medical services. Social examination, with its immense reach and specialized progresses, holds the commitment of getting another period of expectation and preventive medical services, changing the climate of patient consideration and general wellbeing (Ghazal et al. 2022).

2. Impact of Lifestyle Choices on Disease Prediction

The study of how lifestyle choices affect our ability to predict diseases is a significant component of contemporary healthcare research. Dietary propensities, specifically, assume a significant part in forming a singular's wellbeing movement. We can learn more about the nutrients that people consume by looking at their eating habits, which helps improve the accuracy of disease prediction models. Diet is a significant piece of living since it straightforwardly influences who liable to get infections and how rapidly the infections deteriorate. New examinations stress the significance of concentrating on the better places of food decisions, similar to what supplements are eaten and what they mean for various wellbeing results. This top to bottom examination helps find drifts that can be utilized to improve illness forecasts (Djerioui et al. 2020).

Figure 1 shows a total rundown of the relative multitude of significant supplements for wellbeing, including nutrients, minerals, macronutrients, and cell reinforcements, alongside definite depictions of how each one assists the body remain healthy and work appropriately. Not only does this visual guide show where to get each vitamin and how much you should eat each day, but it also explains how vitamins work in the body and what could happen if you don't get enough or enough of them. In addition, the figure shows how various supplements cooperate to help metabolic cycles, resistant capability, mental wellbeing, and illness counteraction. It is a helpful instrument for nutritionists, medical services experts, and individuals who need to eat a solid, adjusted diet to retain sound areas of strength.

Our understanding of the ways in which lifestyle choices affect health grows as we determine whether there is a link between exercise habits and disease risk. Actual activity that you do routinely has been connected to numerous medical advantages, for example, a lower opportunity of getting ongoing infections. It is possible to obtain a more comprehensive picture of a person's health and disease risk by including exercise-related data in disease prediction models.

It is not sufficient to merely examine a person's calorie intake; it is additionally necessary to take a gander at the sorts of things they eat, how nutritious they are, and what they could mean for specific sicknesses. A lower opportunity of coronary illness has been connected to the Mediterranean eating routine, which is loaded with natural products, veggies, and solid fats. This more mind-boggling information on food choices makes it conceivable to make models that can precisely distinguish infections (Choudhary et al. 2023).

Assembling various kinds of information is expected to sort out what dietary patterns mean from the standpoint of an illness. Prescient models work better when they utilize both self-announced food data and goal estimates like biomarkers and hereditary information. Then, more mind-boggling factual methods and AI projects can find little patterns that assist with making a more exact conclusion of infection risk.

The connection between practice propensities and infection risk incorporates various parts of active work, for example, how frequently, how hard, and how lengthy you make it happen. Concentrates on demonstrating the way that even gentle activity

Figure 1. Nutrients for health.

consistently can bring down the possibility of getting sicknesses like diabetes and coronary illness. Adding itemized data regarding exercise propensities to prescient models assists us with understanding what exercise means for well-being in a more mind-boggling manner.

An individual's dietary patterns can be changed to bring down their chance of getting a disorder. Healthcare professionals are able to provide targeted treatments, such as individualized diet advice and lifestyle modifications, by identifying patterns that are associated with an increased risk of disease. The advance toward preventive medical care techniques is in accordance with this proactive methodology, which depends on nourishment research.

While investigating the connection between practice propensities and illness risk, it is vital to remember that actual work patterns change after some time. Longitudinal examinations that gander at how exercise propensities change over the long haul can provide us with a great deal of information about what they mean for wellbeing after some time. These kinds of timing factors assist with making forecast models that consider how living decisions change after some time.

To utilize dietary investigation to anticipate sicknesses, you must move beyond issues with information rightness and constancy. Utilizing innovations like wearables and portable applications progressively settles on simpler-to-follow food decisions, which works on the nature of the information that can be examined. These innovations in the healthcare systems work fairly with the data, and it is required to have constant updates and training to improve efficiency.

Hereditary and outer elements cause the connection between what is eaten and the illness to get extremely convoluted. By adding genome information to prescient models, a more customized technique can be utilized to find individuals who might be hereditarily bound to encounter specific wellbeing chances connected to their food. This degree of exactness makes sickness expectation models better at coordinating cantered counteraction steps.

Trends that go beyond heart health are discovered when exercise habits are examined in terms of disease risk. For instance, the impact of exercise on mental health demonstrates how lifestyle choices affect a person as a whole. Adding psychological wellness elements to forecast models expands the quantity of factors that can be checked out and gives a full image of what living decisions mean for general wellbeing (Hegde et al. 2022).

Differences in health outcomes can be identified by examining the eating and exercise routines of individuals belonging to particular groups. Things like people's culture, socioeconomic status, and access to good food influence how they choose to live. Changing illness forecast models to consider these distinctions ensures that arrangements are receptive to various societies and meet the particular necessities of these gatherings.

Accuracy medication is by the possibility that checking dietary patterns and gym routine schedules can assist with foreseeing sicknesses. Medical care therapies can be customized to target explicit gamble factors by considering how various individuals' living decisions are, making precautionary steps more successful. This customized strategy is a major step in medical services toward additional engaged and patient-focused approaches.

When combined with conduct information, dietary examination gives us a superior image of the mental and social pieces of food decisions. Profound eating, eating designs brought about by pressure, and social factors make dietary patterns more muddled. Given the perplexing connection between the brain and food choices, remembering mental elements for models that attempt to anticipate infections and predictions more clearly.

Learning techniques for the investigation of dietary patterns and workout schedules makes models more exact at foreseeing what will occur. These calculations can find complex connections and connections that don't follow a straight line in different datasets. They can likewise find drifts that are difficult to see with customary measurable strategies. The reception of cutting-edge PC strategies adds to the improvement of sickness gauge models.

Because habits can change over time, the timing of living choices is very important for disease forecasting. More precise forecast models are made with the assistance of long-haul concentrates on that track changes in eating and exercise propensities. In understanding how living decisions change makes infection figure frameworks more adaptable and exact (Deng et al. 2022).

Medical care laborers, information researchers, and specialists need to cooperate to see dietary patterns and exercise routine schedules to foresee infections. Cooperating across various fields ensures that prescient models consider a great many elements, from clinical signs to unobtrusive changes in conduct. This joint technique assists individuals with getting a full image of the convoluted connection between living decisions and wellbeing results.

Dietary examination is significant in light of the fact that it can assist with directing centered activities that stay away from disease. By finding explicit dietary patterns that are connected to a higher gamble of infection, medical care experts can make customized arrangements, like contributing healthful direction and preparing programs. The objective of these projects is to give individuals the data they need to settle on better food choices, which will bring down the opportunity of medical conditions.

To investigate the connection between practice propensities and sickness risk, the external factors that influence the activity propensities of individuals must be pondered upon. How often people exercise can be affected by where they live, where they work, and how easy it is to get to parks and other places to relax. Adding outer variables to models that anticipate illnesses assists us with looking further into what pursuits individuals settle on living choices and what those choices mean for their wellbeing.

Figure 2 shows an outwardly complete gander at the perplexing connection between practice propensities and sickness risk. It shows how exercise colossally affects numerous wellbeing results. The effects of aerobic, resistance, and flexibility training on the risk of developing chronic diseases like heart disease, diabetes, obesity, and some cancers are depicted in this figure. It does this by utilizing intriguing designs and information representations. The figure shows what normal activity means for

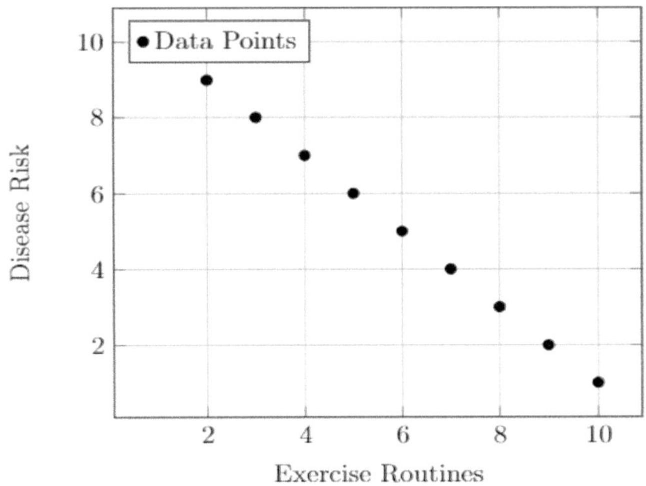

Figure 2. Exploring the correlation between exercise routines and disease risk.

biochemical cycles, like aggravation, insulin responsiveness, oxidative pressure, and metabolic control. These progressions assist with staying away from illness and work on broad well-being. This visual guide not only presents the defense for integrating exercise into public and clinical well-being programs more grounded, but it also demonstrates how carrying on with a functioning way of life can lessen the number of infections individuals get and work on their satisfaction.

3. Activity Patterns and Health Outcomes

The intricate connection between how people behave and their health has emerged as a major focus of contemporary healthcare research. Understanding the impacts of action levels on illness conclusion requires an exhaustive investigation of how different types of activities work. Humble activity to ways of behaving influence a singular's overall wellbeing. Action immensely affects the possibility of getting an infection. Researchers have identified that having daily exercise benefits mental and physical well-being, energy and healthy heart conditions. Concentrating on the nuances of various activity levels is significant for creating prescient models that can accurately sort out how likely an individual is to getting different medical conditions (Hammoud et al. 2023).

With regards to wellbeing impacts, it is similarly vital to quantify inactive propensities. Chronic diseases like obesity, heart disease, and metabolic syndromes have been linked to modern idle lifestyles, which involve sitting for long periods of time and not getting much exercise. At the point when you add information on stationary propensities to models that foresee illnesses, you can find out what significant time frames of dormancy can mean for an individual's wellbeing.

To comprehend what exercise levels mean for infection conjecture, it is necessary to ponder upon the various types of actual work individuals do consistently. From arranged exercises to spontaneous moves, each kind of activity diversely affects overall wellbeing. Counting this variety makes movement design examination more definite, which thus makes illness forecast models more exact.

Adding actual activity to models that anticipate infections creates a variable element that individuals can use for preventive medical care. For instance, finding examples of movement that are connected to bringing down infection gambles allows us to target medicines. This is similar to custom activity orders and conduct intercessions, that are meant to work on broad wellbeing and lower conceivable wellbeing chances (Nguyen et al. 2022).

To precisely quantify inactive propensities, it is important to do more than count how much time is spent in stationary positions. Examples of stationary ways of behaving are extensive stretches of sitting and different kinds of inert activities. These can tell us a great deal about what they mean for wellbeing. The comprehensive study contributes to the development of forecasting models that are able to detect even minute shifts in sitting patterns and the ways in which they impact health outcomes.

Levels of activity have influences on illness gauges that go past actual wellbeing and incorporate emotional well-being also. Consistent actual activity has been connected to lower levels of psychological wellness issues like trouble and stress. Adding emotional well-being variables to gauging models assists us in studying the general advantages of specific work-out schedules.

When taking a gander in foreseeing infections, movement designs show ways of behaving that go past what every individual decides to do. Movement levels are impacted by friendly and normal variables, which thus influence the strength of the entire society. Including data regarding social cooperations and ecological elements assists us with understanding the greater variables that influence practice propensities and what they mean for general wellbeing.

Sorting out how long individuals spend sitting down is particularly significant now that more individuals telecommute and invest more energy before a screen. Sitting for long periods of time and doing sitting exercises before a screen have become normal, which is terrible for wellbeing. Sickness forecast models that consider the subtleties of inactive ways of behaving in the advanced age can assist us with understanding the wellbeing dangers that accompany current lives (Wang et al. 2022).

Assembling information from various sources is expected to help us comprehend how exercise levels influence foreseeing illness. Movements can be monitored in real-time with smart sensors, wearable technology, and smartphone apps. Utilizing these advancements offers consistent information assortment, working on the exactness and speed of data put into sickness expectation models.

We must carefully examine how long-term inactivity affects other aspects of our lifestyle in order to accurately measure idle habits. Things like how much sleep individuals get, what they eat, and how stressed they are is linked to sedentary lifestyles. These factors influence the overall wellbeing of an individual. Remembering these connected elements for sickness expectation models will assist us in getting a full image of what stationary propensities mean for wellbeing.

The effects of exercise levels on disease forecast become clearer when you look at how physical activity changes with age. Various sorts of activity diversely affect wellbeing in youngsters, grown-ups, and more established individuals. Accordingly, adjusting disease prediction models to take into account activity patterns that alter with age ensures that treatments are targeted appropriately across a lifespan.

To quantify inactive propensities, we want to find gatherings that are particularly prone to being wounded by being dormant. Individuals who work at a work area the entire day or who don't have simple admittance to public recreation regions might be bound to do things that hold them back from moving for extensive stretches of time. Creating custom fitted medicines for these gatherings needs a point-by-point information on their particular activity examples and wellbeing gambles.

At the point when action patterns are taken a gander at to attempt to foresee illnesses, they show how wellbeing results change over the long haul. Concentrates that track changes in practice levels over the long run, called longitudinal examinations, assist with making prescient models that consider how conduct changes after some time. Accordingly, this time study makes disease forecast systems more adaptable and trustworthy.

To measure idle propensities, it is not adequate to simply figure out how long individuals do not move; it is additionally important to understand why they remain stationary. Screen time, whether it is for work or for no particular reason surely affects wellbeing. Sickness forecast models that can differentiate between various kinds of dormant ways of behaving assist us with more deeply studying what they mean for wellbeing.

To comprehend what exercise levels mean for sickness figure, we need to take a gander at the issue from a world perspective that considers how individuals in various societies do proactive tasks. People's levels of physical activity are influenced by social and cultural norms, which in turn affects their health. Sickness expectation models that consider social contrasts give more exact evaluations of wellbeing gambles in specific gatherings (Vidya and Sasikumar 2022).

One important part of dealing with issues that come with living in a modern, tech-driven world is quantifying idle behaviors. Screens and occupations that don't need a lot of movement are making individuals less dynamic. In light of today's challenges, models that predict diseases based on sedentary behavior, aid in the development of prevention strategies.

Technology-based treatments could help people live better lives, if activity trends are examined to see whether they can help identify diseases. Individual activity applications, shrewd tech, and game-like components actually urge individuals to participate in actual work. The connection between technology and preventative healthcare strengthens when these technological advancements are incorporated into disease prediction models.

We must address the obstacles that prevent people from obtaining the tools they need to become more active before we can measure idle behavior. Different development patterns are brought about by things like metropolitan preparation, admittance to green regions, and monetary status. The factors that influence activity levels and health outcomes are better understood by disease prediction models that take these environmental determinants into account.

4. Social Interactions and Disease Prognosis

The estimation of social relations in sickness expectation is a developing area of examination that sees the tremendous impact of social connections on wellbeing results. To completely comprehend the job of social connections in recognizing sicknesses, we need to take a gander at how an individual's connections, informal communities, and contribution locally influence their overall wellbeing.

It is becoming increasingly apparent that social ties can influence the course of a disease. Concentrates show that individuals who have a great deal of social help have better psychological well-being, safe frameworks that work better, and general wellbeing. Concentrating on the nuances of social connections can assist with making more exact expectation models that can sort out how likely an individual is to get different medical conditions (Zhang et al. 2023).

The impact of social conduct on mental and actual wellbeing are mind boggling and go past what people pick. Relationships that are good have shown to be of help with mental health issues, heart health, and stress reduction. By adding data about friendly propensities to models that foresee illnesses, we can find out about what different social circumstances can mean for wellbeing.

While taking a gander at how social associations can assist with foreseeing illnesses, it is essential to comprehend the various kinds of organizations and the nature of connections within them. Each sort of friendly connection, from family bonds to kinships to bunch enrolments, is unique and helps an individual's wellbeing in its own specific manner. The inclusion of this variety facilitates a deeper investigation of social relationships, which in turn improves the accuracy of disease prediction models.

At the point when the social way of behaving is added to models that foresee infection, it turns into an inconsistent variable that patients and medical care laborers can use to assist with keeping away from sickness. For instance, finding examples of social associations connected to better well-being results makes it conceivable to concentrate on medicines like local area-based programs and emotional wellness support, which means working on broad prosperity and lower conceivable well-being gambles (Tang et al. 2022).

It takes an entire individual approach to managing social connections to completely comprehend what social conduct means for mental and actual well-being. Positive connections assist with helping people be versatile and solid, and regrettable social propensities, such as being separated from everyone else or in a poisonous relationship, have been connected to terrible well-being results. Adding these various elements to models of sickness expectation will assist us in better understanding the muddled connection between friendly activities and wellbeing.

At the point when you take a gander at social connections with regard to illness forecast, you can see examples of conduct that go past individual decisions and affect entire networks. Wellbeing results are enormously impacted by friendly drivers of wellbeing, which incorporate things like monetary position and admittance to medical care. Accordingly, incorporating information about social processes and causes into disease prediction models enhances our comprehension of how broader societal factors influence health.

Examining how social ties can assist in the identification of illnesses is especially helpful when attempting to determine how mental health outcomes occur. It has been discovered that social support helps prevent mental health issues because it creates a sense of connection and improves emotional health. Adding psychological wellness variables to prescient models assists us with looking further into the many advantages of having great associations with others.

At the point when the social way of behaving is taken a gander at for infection expectation, it can assist with finding gatherings that are at a high gamble for terrible well-being results. Some groups may be more susceptible to certain health risks, such as those who live in social isolation or have to deal with structural inequality. To do extraordinary projects to assist these gatherings, we need to completely get a handle on their one-of-a-kind social circumstances and well-being chances.

Assembling information from various sources is expected to sort out how social connections can assist with recognizing sicknesses. Reviews, informal organization examination, and local area wellbeing studies show how social connections are organized and how they work. Utilizing these different information sources considers improving the exactness and handiness of data put into illness expectation models.

To sort out what social conduct means for mental and actual wellbeing, we want to take a gander at how social connections differ across societies. Social standards, customs, and local area norms influence the sort and nature of social connections, which thusly influences wellbeing results for many gatherings. Sickness forecast models that consider culture contrasts give more accurate gauges of wellbeing gambles in specific gatherings.

At the point when you take a gander at social connections for illness expectations, you can perceive how innovation-driven medicines could work on great social propensities. Computerized devices like interpersonal organizations, medical care administrations, and advanced stages can assist people in making new companions, particularly in places that are far away. Adding these specialized changes to models that foresee illness prompts a positive connection between preventive medical care and innovation.

Since social connections can assist with anticipating illnesses, it is necessary to fix the fact that certain individuals do not have equivalent admittance to social instruments that influence wellbeing. Wellbeing results differ contingent upon financial elements, the offices of the local area, and the arrangement of social encouragement groups of people. Illness forecast models that consider these social drivers provide us with a full image of the factors that influence social connections and what they mean for wellbeing results.

At the point when social way of behaving is taken a gander at to foresee an illness, it shows how wellbeing impacts change after some time. Longitudinal investigations that glance at how social ties change after some time assist with incorporating prescient models that consider how conduct changes over the long run. This time study works on the adaptability and soundness of sickness forecast techniques (Iqbal 2024).

To sort out what social conduct means for mental and actual wellbeing, it must be understood comprehend that social connections are continuously evolving. Wellbeing ways are impacted by changes in informal communities, large occasions throughout everyday life, and advances. Models for anticipating sicknesses that consider these progressions in friendly conduct after some time assist us with understanding how social cycles and wellbeing impacts change over the long haul and how perplexing it is.

To completely comprehend how social connections can assist with recognizing sicknesses, we want to take a gander at them from a world perspective that considers how individuals cooperate with one another in various societies. Social standards, customs, and local area norms influence the sort and nature of social connections, which in turn influence wellbeing results for a great many gatherings. Therefore, health risks of certain groups are more accurately estimated by disease prediction models that account for cultural differences.

When social contacts are taken a gander at as far as sickness expectations are concerned, they show how centered general wellbeing measures could function. To care for specific medical conditions, local area-based projects, social endorsing programs, and psychological wellness mindfulness endeavors can utilize what we know about how individuals act in friendly circumstances. Placing these medicines into models anticipating illness results prompts more complete and powerful avoidance plans.

While taking a gander at how social connections can assist with distinguishing illnesses, it is essential that social propensities and wellbeing results are connected in the two bearings. Wellbeing is impacted by friendly ties, yet medical conditions can likewise influence how individuals' partner with others. Illness expectation models that recognize this powerful communication offer a more complete information on the intricate connection among wellbeing and social variables.

While taking a gander at social ways of behaving for illness expectation, it is seen to be very critical to be engaged with the local area and to give individual's power. People's mental and physical health have shown to improve in communities where people get along with each other. We can learn more about how social behaviors affect health as a whole by using disease prognosis models that include indicators of community activity.

Figure 3 shows the complicated connection between friendly associations and infection standpoints. It is a fascinating visual portrayal of how social associations, encouraging groups of people, and local area contribution hugely affect wellbeing results. This image shows the numerous ways that social collaborations can change the result of an illness. It likewise shows how having great associations with others can work on psychological wellness and assist with managing ailments. Likewise, it shows how social help, feeling that you have a place, and having significant connections might assist with actual opposition, bringing down pressure, changing the insusceptible framework, and general adapting strategies, all of which can influence how an infection advances and how well you recuperate. This visual guide shows the confounded trap of social determinants and what they mean for well-being. It not only shows that it is essential to help social associations in medical services mediations, but it also upholds an entirely individual way to deal with sickness. The

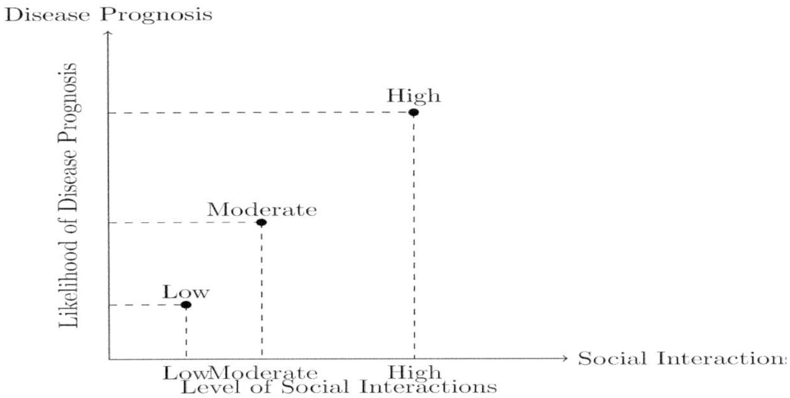

Figure 3. Social interactions in disease prognosis.

executives consider social collaboration to be a significant job in building strength and prosperity.

5. Techniques Utilizing Behavioral Analysis for Disease Prediction

Understanding the strategies utilized in conduct examination is a significant piece of figuring out how to foresee sicknesses. This outline covers a great many devices and techniques that utilize conduct information to make forecast models better. At the point when social examination strategies are consolidated with different techniques, similar to standard survey techniques and state of the art simulated intelligence applications, it becomes easier to comprehend and foresee wellbeing results.

Trend setting innovations are frequently utilized with conduct exploration to assist with recognizing infections. Wearable tech, applications, and brilliant contraptions generally gather constant information on how individuals act, providing us with a great deal of data about our rest, practice levels, and way of life decisions. These innovations consider steady checking, which provides us with a full and itemized image of how individuals act, which is utilized to further develop estimating models.

Utilizing machine learning techniques is a significant component of using behavioral research to predict diseases. These intricate models can take a gander at extremely huge datasets and find patterns and associations that may be difficult to come by with customary techniques. The power of artificial intelligence is used in machine learning to boost a models' accuracy and predictive abilities. This makes it conceivable to anticipate illnesses all the more precisely over the long haul.

Conduct investigation is additionally used to distinguish illnesses through interpersonal organization examination. In order to identify potential leaders, behavior clusters, and overall trends, this strategy examines how individuals join and interact in social networks. Prescient models get a more complete picture when they comprehend what social variables mean for individuals' wellbeing propensities.

This is particularly obvious while taking a gander at irresistible illnesses and local area wellbeing.

An important tool in behavioral research is natural language processing (NLP), which looks for patterns in words in text data. NLP can be utilized on different sources, like electronic wellbeing information, web-based entertainment, and online networks, to get significant information about the psychological wellness individuals, close to home prosperity, and conceivable wellbeing gambles. Adding NLP to models that anticipate infections assists us with delving deeper into what mental variables mean for wellbeing results.

Contextual analyses are strong instances of how conduct bits of knowledge can be utilized to foresee infections precisely. Wearable innovation was utilized in one review to show a connection between changes in actual activity propensities and the beginning of persistent illnesses. By adding these conduct elements to expectation models, medical issues could be tracked down right on time and treated as soon as possible.

In a fascinating case study, a machine learning program examined long-term data on lifestyle choices and discovered minute behavioral shifts linked to the expansion of cardiovascular diseases. It was simple for the program to find patterns in individuals' dietary patterns, work-out schedules, and rest quality, which prompted more precise expectations of their wellbeing ways and the production of custom avoidance plans.

It has been possible to anticipate how irresistible sicknesses will spread in networks utilizing interpersonal organization examination. Analysts found notable individuals who influence the spread of sickness by following social associations and collaborations. This social review made it more straightforward for centered general wellbeing measures to stop breakouts and diminish their impacts on local area wellbeing.

Normal language handling has been demonstrated to be helpful for anticipating psychological wellness. By looking at posts on social media and online groups, researchers discovered language patterns linked to depression symptoms. Specialists had the option to precisely foresee emotional well-being results by involving these conduct discoveries in prescient models. This shows the commitment for innovation-based arrangements in emotional well-being following.

Conduct investigation strategies aren't simply used to concentrate on how individuals act; they can likewise be utilized to concentrate on how individuals communicate in networks. Interpersonal organization examination was utilized to examine what local area-based measures meant for well-being brought about in a contextual investigation. To figure out how well-conducted medicines worked on the strength of the entire local area, scientists checked out changes in friendly associations and contributions.

An excellent illustration of this is when behavioral analysis and machine learning were combined to predict the progression of neurological diseases. The model found early indications of illness advancement by seeing little changes in discourse designs, facial responses, and day to day exercises. This multi-faceted investigation of conduct helped think of protection medical care plans for individuals who are in danger.

A concentrate on way of life related tumors, demonstrated the way that various sorts of conduct explorations can be utilized together. Specialists made a full forecast model by blending information from shrewd tech, interpersonal organization trades, and regular language handling of wellbeing related discussions. This multidisciplinary procedure worked on the exactness of malignant growth risk expectations, demonstrating the way that different social examination techniques can cooperate (Gupta et al. 2023).

A contextual investigation of that pre-owned informal community examination demonstrated how it may be very well utilized to foresee how psychological wellness brings about school circumstances. Scientists tracked down significant variables that influence emotional wellness by making an image of the social ties between understudies. Zeroing in on activities in view of the consequences of informal organization examination made a supportive group environment and further developed emotional well-being results.

Natural language processing was used in a case study about efforts to improve public health. By reading what people were saying online about health treatments, researchers discovered a lot about how people feel about them and how involved they are. This investigation of conduct further developed specialized techniques, which made general well-being attempts more effective and urged individuals to do things.

In a notable contextual analysis, AI techniques were utilized to take a gander at electronic wellbeing records utilizing social examination to foresee sicknesses. The model had the option to foresee sickness issues and deterioration by finding patterns in how patients acted, how well they took their drugs, and how frequently they utilized medical care administrations. This utilization of conduct experiences in clinical circumstances showed that customized medical care therapies in view of individual ways of behaving are conceivable.

Conduct research is a vital piece of sorting out what outside factors mean for wellbeing and making expectations about those impacts. Informal organization examination was utilized for a situation study to take a gander at how networks manage ecological issues. Scientists found out about the networks' capacity to adjust by following how individuals converse with one another and what they do collectively. It prompted the improvement of techniques for bringing down the wellbeing gambles with that accompany changes in the climate.

Informal organization examination and normal language handling were both utilized for a situation learning about psychological well-being. Scientists found notable individuals and language designs connected to great outcomes by seeing how individuals in emotional wellness support bunches conversed with one another on the web. This technique, from various fields, provided profound information on what social cooperations mean for psychological well-being brings about virtual settings.

Social exploration has been extremely useful in sorting out what general wellbeing estimates will mean for individuals. In one contextual investigation, informal community examination was utilized to take a gander at how wellbeing data spreads in gatherings. To disclose wellbeing specialized techniques more fruitful and improve individuals' wellbeing designs, specialists investigated how data spreads and who the idea chiefs are.

Wearable innovation and AI were utilized for a situation study to demonstrate the way that living decisions can assist with distinguishing illnesses. The review found individual gamble factors for constant illnesses by monitoring individuals' eating, dozing, and practice propensities over the long run. Focused actions were made possible by incorporating these behavioral insights into predictive models, which resulted in more individualized plans for health promotion and disease prevention.

The techniques that utilize conduct examination to foresee infections show an adaptable and multidisciplinary perspective on results. These strategies give us valuable data. They incorporate state of the art advances like savvy tech and AI, as well as strategies that can imagine interpersonal organization examination and regular language handling. Promoting contextual investigations is another way that conduct experiences can be utilized effectively. They demonstrate the way that customized and zeroed in arrangements can be founded on both individual and gathering conduct.

6. Challenges and Future Prospects

Despite the fact that there has been a positive step in the right direction in utilizing social examination to foresee illnesses, there are still issues that should be painstakingly and thoroughly examined—beginning with the fact that social insights could be one-sided. Determination predisposition in concentrate on subjects, contrasts in sociocultural foundation, and contrasts in admittance to innovation can all prompt slanted depictions, which can make forecast models less valuable.

It is vital to ponder upon morals while utilizing social examination to foresee illnesses. Individuals are stressed over their protection as gathering data about their web-based propensities is common. Tracking down the right blend between getting helpful data and safeguarding individuals on the whole correct to security is difficult to track. To take care of these ethical issues, clear principles about how to utilize information must be framed that make provisions for the need for unmistakable authorization from individuals and incorporate solid techniques for concealing information.

A major issue in social examination is that AI models are difficult to comprehend. Despite the fact that mind boggling calculations can make great speculations, it is difficult to perceive how choices are made because that the calculations are difficult to understand. Ensuring that models can be perceived is vital, particularly in medical services, where clear direction is expected to fabricate trust among patients and medical services laborers.

Attempting to think about what infections individuals will get is hard when social patterns change over the long haul. Innately, conduct information changes over the long run because of a great many variables, including life altering situations, changes in the seasons, and changes in the public eye. To keep sickness expectations valuable and right over the long haul, figure models should be changed to take these progressions in time into account.

Another issue is that social examination techniques should have the option to be utilized on a great many individuals. Contrasts in culture, language, and financial status can hugely affect how all-around conjecture models work overall. These natural variables should be painstakingly and thoroughly examined to ensure that social examination is comprehensive and helpful for ethnic gatherings.

There is a lot of promise in using behavioral data to predict diseases in the future to solve current issues. The development of explainable artificial intelligence (XAI) aims to simplify machine learning models. By making confounded calculations more noticeable and comprehended, XAI aids in incorporating trust in expectation models and supports their joining into medical care dynamic cycles.

Utilizing more than one sort of information together could assist with making infection expectation models better. It is possible to create a more complete picture of health risks by combining clinical data, behavioral data, DNA data, external factors, and other data. By cooperating across various fields of information, science could prompt a more thorough comprehension of what the many elements mean for illness results and how they influence one another.

The advancement of wearable technology and sensor capabilities is encouraging for the expansion of behavioral data collection. Wearable tech that is continually being followed can gather information progressively, giving a more precise image of how individuals act over the course of the day. Data become even more detailed as a result of recent advancements in monitoring technologies, revealing minute behavioral shifts that may precede health issues.

In the future, it is possible that personalized and precision medicine will be used to identify diseases more frequently. Prescient models that are custom fitted to every individual's characteristic, tastes, and responses to medicines can make anticipation strategies significantly more exact and helpful. Personalized treatments based on behavioral insights might lead to health solutions that are more focused and efficient.

A vital issue to contemplate is creating standard models for moral conduct research. To deal with the moral issues about involving social information in medical services, it is critical to set clear principles for information control, teach authorization, and ensure security. The advantages of conducting an investigation can be acquired without disregarding individuals' freedoms the same length as friendly accepted procedures are followed.

For future advancement, it is essential to explore what conducting an examination in illness conjecture means for society. It is essential to formulate equitable and inclusive plans for all parties to consider the larger effects of utilizing prediction models in healthcare situations, such as how this could result in differences in access and outcomes. While contemplating morals, we shouldn't just ponder upon the security of individuals and what conducting research means for society and networks overall.

Methods for natural language processing, or NLP, could benefit behavioral analysis in new ways. NLP can be improved to all the more likely get on minor language signs that show emotional wellness states, which will take into account more exact forecasts. The field of conducting examinations in psychological wellness figures is continuously changing because NLP calculations are improving and can be utilized in a more extensive scope of text-based sources.

To ensure that prescient models are dependable and can be utilized in different circumstances, solid approval and replication methods should be made. Discoveries can be rehashed assuming strategies are obviously made sense of, tests are shared, and specialists from various schools cooperate. Setting up a culture of severe assessment supports the dependability of conduct examination in foreseeing illnesses and fabricates trust in established researchers and then some.

For the social examination to be utilized in illness recognition, later on, adopting a proactive strategy for information security issues is critical. Utilizing solid safety efforts, encryption strategies, and safe information-sharing methodology can assist with keeping private conduct information from being taken or abused. Ensuring that individual well-being data is retained and private is a vital part of conducting examination devices morally.

The area of examining infection conclusion can push ahead an incredible arrangement through worldwide organizations and information-sharing projects. It is easier to combine various datasets when international standards for data exchange, methods coordination, and shared libraries are established. This joint technique speeds up the quest for general patterns in social information, which prompts prescient models that can be utilized all the more broadly.

Focusing on client focused plan ideas is turning out to be increasingly more significant as innovation changes conduct examination apparatuses so that they function admirably in medical care. Making these apparatuses simple to utilize, accessible, and receptive to various societies makes them bound to be utilized and more valuable. In order for behavioral analysis tools to be more effective in real-world healthcare settings, end users—such as individuals and healthcare professionals—should be involved in their creation and development.

Examining how behavioral research could use blockchain technology to predict diseases is one way of looking at the future. Since blockchain is autonomous and safe, it can assist with information protection issues and give individuals more command over their wellbeing data. Managing and disseminating behavioral data in the healthcare industry might be made simpler and more stable with the help of blockchain solutions.

Zeroing in on reasonableness might assist social examination with pushing ahead both in AI models and in how results are imparted to individuals and medical services laborers. Ensuring that assumptions are made sense in a clear and usable manner works on the handiness of conducting examinations to make savvy decisions in both precaution and crisis medical care settings.

Placing social examination into medical care therapies that occur progressively has a great deal of potential. Telehealth, versatile applications, and brilliant well-being devices have made considerable progress, making it conceivable to integrate social information into custom medical services plans effectively. This proactive and adaptable strategy allows medicines to be given in the exact timing with precision and with professional conduct standards, which works on the general proficiency of preventive medical services.

7. Conclusion

Investigating utilizing conduct examination to distinguish infections has shown us a world loaded with open doors, issues, and moral inquiries. The main findings indicate that behavioral insights have the potential to completely alter healthcare's future. They do this by taking a gander at a great many elements, like living decisions, social collaborations, and new instruments and strategies. This is significant in light of the fact that social examination can work on prescient models, yet it can likewise begin customized and precise medical care. The main findings indicate that using disease forecast models with behavioral analysis provides us with a more comprehensive picture of how the health of individuals and groups changes over time. When it comes to predicting how a disease will progress, lifestyle choices, social connections, and exercise routines take on a significant role. In addition, the applied approaches, such as natural language processing, machine learning, and social network analysis, demonstrate that behavioral analysis is multidisciplinary and provides a comprehensive picture of health.

These outcomes are significant for something beyond research since they show that involving conduct perceptions in healthcare is significant. Focused treatments, preventative measures, and more individualized healthcare is made possible through the prediction of diseases based on behavioral trends. Doctors are able to assist patients sooner and more effectively when they are able to identify minute behavioral shifts that occur before the onset of health issues, which ultimately results in improved health outcomes over time. As we manage imperfections, social issues, and how we conduct information changes over the long haul, obviously, the eventual fate of medical care lies in tolerating the advantages and fixing the issues of social examination. To make the future of behavioral analysis in healthcare more accountable and accessible to everyone, the expansion of AI that can be explained, user-centered design, and global partnerships are all very important. We can make a better approach for pondering upon medical services that puts individuals' wellbeing first, empowers precaution care, and perceives the changing force of social examination in sickness estimate by perceiving the significance of conduct perceptions.

References

Abdeen, M., Ahmed, M. H., Seliem, H., El-Nainay, M. and Sheltami, T. R. (2021). Improving the performance of ambulance emergency service using smart health systems. 2021 IEEE/ACM Conference on Connected Health: Applications, Systems and Engineering Technologies (CHASE).

Alahmadi, A. and Soh, B. (2011). A smart approach towards a mobile e-health monitoring system architecture. 2011 International Conference on Research and Innovation in Information Systems.

Choudhary, C., Nagra, L. S., Das, P., Singh, J. and Jamwal, S. S. (2023). Optimized ensemble machine learning model for chronic kidney disease prediction. 2023 International Conference on Computing, Communication, and Intelligent Systems (ICCCIS).

Deng, Y., Qin, H.-Y., Zhou, Y.-Y., Liu, H.-H., Jiang, Y., Liu, J.-P. et al. (2022). Artificial intelligence applications in pathological diagnosis of gastric cancer. Heliyon 8(12): e12431. https://doi.org/10.1016/j.heliyon.2022.e12431.

Djerioui, M., Brik, Y., Ladjal, M. and Attallah, B. (2020). Heart Disease prediction using MLP and LSTM models. 2020 International Conference on Electrical Engineering (ICEE).

Dou, N., Deitch, R., Kowalski, A. J., Kuhn, A., Lane, H., Parker, E. A. et al. (2024). Studying the impact of COVID-19 mitigation policies on childhood obesity, health behaviors, and disparities in an observational cohort: Protocol for the COVID-19 Family Study. Contemporary Clinical Trials 136(107408): 107408. https://doi.org/10.1016/j.cct.2023.107408.

Geeitha, S., Karthikeyan, P., Aravinth, S., Nachiappan, P. and Jagadeesh, S. (2023). Contagious disease prediction using random forest algorithm interpolated with fuzzy model. 2023 International Conference on Sustainable Computing and Data Communication Systems (ICSCDS).

Ghazal, T. M., Rehman, A. U., Saleem, M., Ahmad, M., Ahmad, S. and Mehmood, F. (2022). Intelligent model to predict early liver disease using machine learning technique. 2022 International Conference on Business Analytics for Technology and Security (ICBATS).

Gupta, R., Kumari, S., Senapati, A., Ambasta, R. K. and Kumar, P. (2023). New era of artificial intelligence and machine learning-based detection, diagnosis, and therapeutics in Parkinson's disease. Ageing Research Reviews 102013.

Hammoud, M., Kovalenko, E., Somov, A., Bril, E. and Baldycheva, A. (2023). Deep learning framework for neurological diseases diagnosis through near-infrared eye video and time series imaging algorithms. Internet of Things 24: 100914.

Hegde, S., Ajila, V., Zhu, W. and Zeng, C. (2022). Artificial intelligence in early diagnosis and prevention of oral cancer. Asia-Pacific Journal of Oncology Nursing 9(12): 100133. https://doi.org/10.1016/j.apjon.2022.100133.

Huang, P.-C., Lin, C.-C., Wang, Y.-H. and Hsieh, H.-J. (2019). Development of health care system based on wearable devices. 2019 Prognostics and System Health Management Conference (PHM-Paris).

Iqbal, J. M. (2024). A novel deep learning approach for early detection of cardiovascular diseases from ECG signals. Medical Engineering & Physics 125: 104111.

Nguyen, D., Nguyen, H., Ong, H., Le, H., Ha, H., Duc, N. T. et al. (2022). Ensemble learning using traditional machine learning and deep neural network for diagnosis of Alzheimer's disease. IBRO Neuroscience Reports 13: 255–263. https://doi.org/10.1016/j.ibneur.2022.08.010.

Rane, A. L. (2018). A survey on intelligent data mining techniques used in heart disease prediction. 2018 3rd International Conference on Computational Systems and Information Technology for Sustainable Solutions (CSITSS).

Shalini, V. B. (2021). Smart health care monitoring system based on internet of things (IOT). 2021 International Conference on Artificial Intelligence and Smart Systems (ICAIS).

Tang, S., Yuan, K. and Chen, L. (2022). Molecular biomarkers, network biomarkers, and dynamic network biomarkers for diagnosis and prediction of rare diseases. Fundamental Research.

Vidya, B. and Sasikumar, P. (2022). Parkinson's disease diagnosis and stage prediction based on gait signal analysis using EMD and CNN–LSTM network. Engineering Applications of Artificial Intelligence 114: 105099.

Wang, C., Xu, T., Yu, W., Li, T., Han, H., Zhang, M. et al. (2022). Early diagnosis of Alzheimer's disease and mild cognitive impairment based on electroencephalography: From the perspective of event related potentials and deep learning. International Journal of Psychophysiology.

Zhang, Y., Wang, Y., Zhang, B., Li, P. and Zhao, Y. (2023). Methods and biomarkers for early detection, prediction, and diagnosis of colorectal cancer. Biomedicine & Pharmacotherapy 163: 114786.

6

Multimodal Approach of Data Processing in Healthcare

1. Introduction

Technological advancements have played a significant role in development and innovation over the past two decades. Technological advancements and innovations have evolved in various sectors across industries. Healthcare is one of the sectors where transformative changes have been pivotal. Enhanced focus and development have revolved around data-driven techniques in healthcare. These applications of innovations are in disease prediction, diagnosis, and patient management. Traditional models and their applications in health care started to be replaced with intelligent tools powered by data. Traditional data analysis and treatment plans depend heavily on clinical observations, patient monitoring, and patient-reported information. While giving valuable and reliable insights, conventional approaches could be more comprehensive in capturing the complexities of modern healthcare needs, but treatment plans take longer to get results and to understand whether the treatment will work. Precision medicine is a concept that got empowered with data analysis and personalized treatment plan preparations and increasingly gained large-scale traction in place. Diverse data sources and their integration into the medical systems help to achieve a more personalized and evidence-based approach with the power of data and mining information from the diverse forms of data.

While the focus is on precision medicine, personalized healthcare is increasingly attracting developments and applications in precise-personal healthcare systems. These applications use refined techniques and recent advancements to analyze and integrate data from diverse sources. Electronic health records (EHRs), genomic data, medical imaging data, and real-time physiological metrics are frequently researched data sources for healthcare systems that help create thorough treatment plans for improved clinical insights. In the traditional treatment and data analysis approach,

DOI: 10.1201/9781032722344-6

doctors heavily depend on information provided by the patients, the manual analysis of the documents presented by patients, and the test results.

The dependency on the data from these sources and manual analysis may delay the diagnosis, mainly when it is hard to identify the appropriate mutations and changes in the biomarkers that go unrecognized. Uncommon patterns and observations are common in medical applications. Manual analysis and interventions are commonly required in various diagnosis processes. These manual interventions are required to give additional information and verify the algorithmic predictions made from patient data. For example, while preparing for a treatment, the patients have to give all pertinent information to the doctors during their inspection. The doctors rely on the information provided by the patient to identify and prepare the treatment plans. Understanding complex patterns and indicators of disease is not often considered when detecting the cause of the problem. The modern multimodal data aggregation and analysis techniques and approaches are designed to provide clinicians with accurate data that helps them with a comprehensive understanding of the patient's health profile.

The multimodal approach to data processing has emerged as a strategy to address the limitations of traditional treatment approaches. The system incorporates heterogeneous data sources relatable to the patient in the multimodal data analysis and treatment plan preparation. This helps in obtaining a holistic view of the patient's health. Intelligent systems with complex data processing and sharing systems help to relate a particular patient with similar cases in the system by giving significant information to make faster decisions. Medical decisions are heavily dependent on data. The evidence from the data drives the decisions of doctors, leading to making healthcare a data-driven decision-making system.

Multimodal data processing involves preparing and analyzing data from multiple heterogeneous sources. This multimodal data processing approach creates a comprehensive information repository on patient profiles, diseases, treatment plans, etc., generating a centralized repository of medical data. Multimodal data processing includes the processing and organizing of data from diverse types, including textual, numerical, and visual modalities. Studying these data from various sources gives a better understanding of the disease and its mechanisms and improves accuracy in diagnosing diseases. Understanding the data from various regions will again enhance the completeness and usability as various mutations are considered in this case.

Combining medical imaging data with clinical reports and genomic profiles has shown an increase in the accuracy of disease classification models. The approach of combining data from different dimensions has shown improvement in treatment plans for oncology and neurology (AlSaad et al. 2024, Duan et al. 2024). There is a requirement of working together to collect and integrate data in order to prepare a centralized data repository. This centralized data repository will be instrumental in studying and understanding hidden patterns. In the time of pandemics like the recent COVID-19 pandemic, the centralized data repository could identify the patterns and the change of mutations in the virus structure could be provided from shared data resources. Hidden patterns and trends in the data are difficult to be detected using the unimodal systems. The multimodal data analysis system has opened the potential for early disease detection and personalized treatment planning.

There is a huge potential and advantages of multimodal data processing, however, significant challenges are present while integrating and using heterogeneous data. Issues including data interoperability, variable quality, and missing information create challenges in easy integration of data from diverse datasets (Putrama et al. 2024). Advanced Artificial Intelligence (AI) algorithms capable of processing complex data, such as Machine Learning (ML) and Deep Learning (DL) algorithms and pipelines help in the processing of data. Multi-task pipelines could combine data from multiple datasets and extract patterns from them. Addressing the challenges of data interoperability and sharing is required to achieve robust and scalable multimodal systems. There should be ethical and policy frameworks required to process and monitor the real-time functioning of these applications.

1.1 Rationale for Multimodal Data Processing in Healthcare

Diseases have varying complexities and structure. In order to understand and analyze the spread and cause of the disease, complex systems and analysis pipelines are required. The limitation in relaying the data from a single data source is that we won't be able to consider the external factors that can also have an impact. Chronic and multifactorial diseases, such as cancer, cardiovascular disorders, and neurodegenerative conditions, involve dynamic interactions between genetic, physiological, and environmental factors. Unimodal data systems and data processing approaches are limited by their dependency on a single data source; these approaches fail to capture the interdependencies, leading to delayed and incomplete diagnoses of diseases and symptoms, risking the lives of the patient. This system takes time and long experimental approaches to conclude treatment plans.

Multimodal data processing offers a system to address the limitations of unimodal and traditional approaches by combining data and insights from various sources. For example, genomic data reveals insights into genetic susceptibility, imaging data provides evidence of anatomical changes, and clinical data offers a record of symptoms and treatment responses. A better patient diagnosis is achieved by integrating the data from EHRs, previous diagnosis history and pattern analysis, medicine intake history, environmental factors, hereditary traits and factors, etc. In neurodegenerative diseases, including Alzheimer's and Parkinson's, the application of multimodal data processing is evident from recent research, and their performance is also significantly promising. Various researches have proved the efficiency of multimodal data processing and analysis for the prognosis and early detection of diseases. Neuroimaging techniques like Computed Tomography (CT), Positron Emission Tomography (PET), and Magnetic resonance imaging (MRI), gives the function and structure of body parts. The information from these sources helps in understanding the disorders (Chakravarthi et al. 2022).

Medical imaging data provide critical information regarding the patient's body part structure and its function. Genetic information helps to identify the possible inherited risk factors. Targeted treatment plans can be designed for patients by combining behavioral and cognitive data. The feature knowledge of the patients can further be enhanced with the knowledge of information from additional sources, such as genetic and neuroimaging data. The data-driven method helps physicians make precise diagnosis decisions. The precise treatment plan guarantees patients receive accurate treatment that is customized to their disease.

Multimodal data processing enables the healthcare system to consider various factors and make decisions regarding patients. These decisions include the patient conditions, medical history, allergic information, genetic traits, etc. Developing an intelligent dynamic system that is driven on data analysis and decisions makes treatment plans more effective. For example, understanding how environmental factors including pollution and air quality impact patient health, can be crucial in making clinical decisions. Clinical, genetic and other relatable data give an accurate picture of a patient's health. The intelligent predictive model improves the diagnostic and treatment accuracy, identifies the possible disease and its cause using a precise diagnostic approach, by understanding the data from the patient's medical and other relatable data.

Adopting and designing a multimodal data processing system for preparing and analyzing patient data represents a shift from traditional, symptom-based, and patient-related diagnosis to a predictive, data-driven approach, where the data is essential. Integrating data from several areas makes a more comprehensive and transparent evaluation of a patient's health possible. AI tools provide extensive information about the health and problems of the patient. Based on these details, doctors can plan treatment. The design of intelligent AI tools is required to make a detailed analysis of a patient's health profile and prepare a precise treatment plan for the patients. Precise treatment plans help doctors and hospitals to offer specialized and targeted interventions for therapies for their patients. Precision medicine improves patient outcomes by accurate treatment of disease, and increases diagnosis accuracy by combining data from several modalities.

1.2 Key Components of Multimodal Data Sources

The key component of a multimodal healthcare system is data. Data analysis gives a comprehensive view of the patient's health, this is done by integrating and analyzing data from several sources. These sources include real-time patient monitoring, imaging data, symptoms, medical treatment history, previous treatment and its details and other factors. Analyzing and gaining insights can be done over a data extraction pipeline that can process and extract relevant information. These data sources can be part of the EHR systems, which constitute as the basis of patient-centered data by storing all the patient information including clinical histories, test results, prescription drugs, and procedure records. An EHR provides organized and historical data that is related to the patient. This helps in assessing and insights to the medical history of the patient and easy access to this information. The analysis of these data sources helps in the design of a system that can effectively understand the disease trajectory, treatment outcomes, and compile a record of patient health. Some of the major components that can be part of the multimodal data or an EHR system for effective patient profile management for precise care of the patients include the following:

Medical Imaging: CT, MRI, and PET are some of the medical imaging techniques that are commonly used for getting detailed information about patient data. These imaging modalities provide the functional and spatial insights of the concerned body parts. They offer precise, non-invasive imaging of interior body structures that are necessary for expert diagnosis, and treatment planning, and also track the illness and treatment progression.

Genomic Data: Genomic data of an individual stores genetic information. Genetic information plays a huge role in deciding the traits of a person. Genomic data decides the biological functions, physical traits, and various other individual characteristics. It also gives information about hereditary facts and relationships. Genomic information can tell the changes that have been caused by the diseases in the individual's health. Analysis of the genomic data will give insights into drug susceptibility, understanding infectious disease mechanisms, and antibiotic resistance of the individual and plan the treatment accordingly.

Real-time physiological data from wearable devices: Real-time data captured from wearable sensors provides deep insights to a patient's health. These systems are currently present in clinics to monitor the patient in a real-time manner. The devices track the heart rate, activity levels, blood oxygen levels, and sleep habits. Along with these devices deployed in the hospitals, patients can link their wearable smart watches and devices to the application to record their health parameters on a regular basis which will also help the doctors with the behavioral data about the patient.

Clinical notes: Clinical notes are documents maintained by doctors and nurses regarding their observations on the patient. This document contains the treatment plan and offers insight on patient care and medication given to the patient at the time of treatment. The document covers the details on disease symptoms and therapies. Observations are recorded as part of clinical notes. One of the main issues with the clinical notes are their unstructured nature and lack of a centralized storage system. Integrating and accessing the clinical notes gives the treatment plan and more information on the patient.

Laboratory and Diagnostic Test results: Various lab tests for blood, microbiology, and pathology reports are often done on patients to learn more about the patient's medical issues. These results measure the parameters that are giving significant information about the patient's current conditions. This information is crucial for illness diagnosis and treatment plan preparation.

Environmental and Lifestyle data: The context and lifestyle information of the patient's health is added by environmental data, such as exposure to contaminants and lifestyle factors. The parameters including stress levels, exercise, and food and benefit the preparation of precise treatment plans. Lifestyle choices, and environmental details in which the patient lived gives significant information on risks and cause of disease.

1.3 Challenges, Opportunities and Future Directions

The creation of intelligent agents that can process, analyze and give important clinical information has several challenges and opportunities. One of the main challenges is data aggregation from multiple sources. Data aggregation from multiple sources needs setting up of standards and the need for complex data processing and integration with existing systems. Multimodal data processing is often faced with the issue of data heterogeneity, integration complexity, and a lack of reliable computational frameworks for data aggregation. It is necessary to overcome the challenges of multimodal data processing by setting standards and applying intelligent multimodal systems in clinical applications. The healthcare sector needs to take advantage of AI tools to ensure adequate care of patients. AI tools could give precise and accurate results in precision medical applications. These systems with the power to access data give more precise diagnoses suggestions and individualized treatment plans. The system's challenges like the lack of data standardization, interoperability, and creating scalable algorithms are to be addressed by setting up standards and policies. Table 1 gives an overview of the challenges of the Multimodal Data processing approach in healthcare systems.

The integration of multimodal data can enhance advancements in precision medicine. Data integration and processing allow the combination of real-time data from wearables, clinical records, imaging, and genetic sources to be combined with historical patient data. Treatment based on the data and its analysis helps to plan individualized treatment. The treatment plans can be prepared precisely to target the disease by considering data from various sources. The precise treatment takes inputs from the patient's genetic information, medical history, and drug resistance. Using technology to analyze the patient and provide timely and individualized care improves patient outcomes and increases accuracy of treatments.

Table 1. Challenges of multimodal approach of data processing in healthcare systems.

Challenge	Description
Data Integration and Interoperability	Integrating data from various sources (EHR, imaging, genomics) is affected by the mismatch in data formats – this makes it a requirement for expensive processing and standardization for compatibility.
Data Quality and Consistency	Data from various sources often lack uniform structure and challenge reliability and integrated analyses.
Privacy and Security Concerns	Handling and combining sensitive health data across multiple sources has privacy and security risks. Data processing should be done by ensuring regulations (Health Insurance Portability and Accountability Act (HIPAA) and the General Data Protection Regulation (GDPR)) and protecting patient privacy.
Computational and Infrastructure Demands	High-dimensional data requires significant computational resources, particularly for real-time processing and advanced ML models.
Ethical and Bias Concerns	ML models planned to be used in healthcare systems need to be comprehensive, and the data from diverse regions should be used to design and train such models to remove bias.

Future developments will focus on creating sophisticated, privacy-conscious, and collaborative frameworks that support scalability and ethical data handling to address the complex demands of multimodal data integration. Federated learning models, for instance, allow healthcare institutions to collaborate by training predictive models on decentralized data. By capturing the interdependencies between various data kinds, advances in AI architectures—particularly in deep learning approaches like graph neural networks and attention-based models—should improve the precision and interpretability of multimodal data fusion. The regulatory bodies should also ensure that multimodal systems uphold ethical standards around transparency, equitable access, and unbiased algorithmic decisions.

As healthcare increasingly moves towards data-informed practices, clinical decision-making is evolving to incorporate and depend on data-driven approaches. Thanks to multimodal data, clinicians can use local and international health databases to compare a patient's health condition with comparable instances. This comparison can promote the creation of individualized treatment plans, shorten the time of therapy, and increase diagnostic accuracy. The various methodologies and applications of multimodal data processing in healthcare, specifically focusing on disease prediction, are also analyzed, and their applications are studied. The chapter discusses the integration techniques for combining diverse data types, reviews state-of-the-art algorithms developed for multimodal fusion, and evaluates the impact of these approaches on predictive performance by analyzing the role of multimodal systems in precision medicine and outlining future directions for research and implementation in clinical practice.

2. Heterogeneous Data Integration for Disease Prediction

Data from multiple sources must be integrated into the system to improve disease prediction. The data sources include genetic data, medical imaging, EHRs, real-time physiological signals, and many others related to the observations that are to be made on the patient. The application of AI helps physicians understand and combine information in various data formats. Data integration and processing make it easy for clinicians to make decisions based on patient profiles. This helps with issues that can arise due to clinical, genetic, and environmental interactions that can be vulnerable for patients. Intelligent algorithms can analyze data and can find complex patterns from the medical data. The algorithms help in processing, integrating the structured and unstructured data, and improve the accuracy of the prediction models.

Effective data integration strategies and processing algorithms are required to process and extract features from the data. ML algorithms for dimensionality reduction help in the processing and in extracting the relevant data for processing and modelling. The data fusion technique works on combining data from multiple sources, enabling more accurate disease prediction. Data integration has challenges due to the diverse sources from which the data are collected for processing and analysis. To handle the issues of data interoperability, quality, and missing information, strategic planning and careful integration need to be created. These steps help in the design of intelligent solutions that support multimodal data processing and handling for the development of intelligent solutions in clinical applications.

2.1 Techniques for Aggregating Structured and Unstructured Healthcare Data

Building medical systems that are effective for predicting and assisting doctors regarding the medical recommendations requires analyzing and processing of data from diverse sources, that are structured and unstructured. Data sources like EHRs, Test findings, and Vital signs monitoring are examples of structured data that are usually well-organized and easy to analyze using ML methods. The unstructured data, gathered from the sources like clinical notes, medical imaging data, clinical test reports, patient narratives, requires advanced AI algorithms and processing approaches to extract meaningful information (Scheurwegs et al. 2016). Data fusion, mapping and analysis are some of the frequently applied approaches to align these inputs into a common dimension that makes the processing easy and effective. Creating a centralized data warehouse with alignment and mapping of the diverse data, helps to design a system that can be queried to obtain real-time data and accurate patient information. This medical data storage helps to focus on designing intelligent algorithms and process pipelines that can work on the data repository and give recommendations and predictions.

Figure 1 shows the diverse data source and the formation of a centralized data warehouse. The central data warehouse has the processing power for the data aggregation from various data sources. This centralized data repository helps to

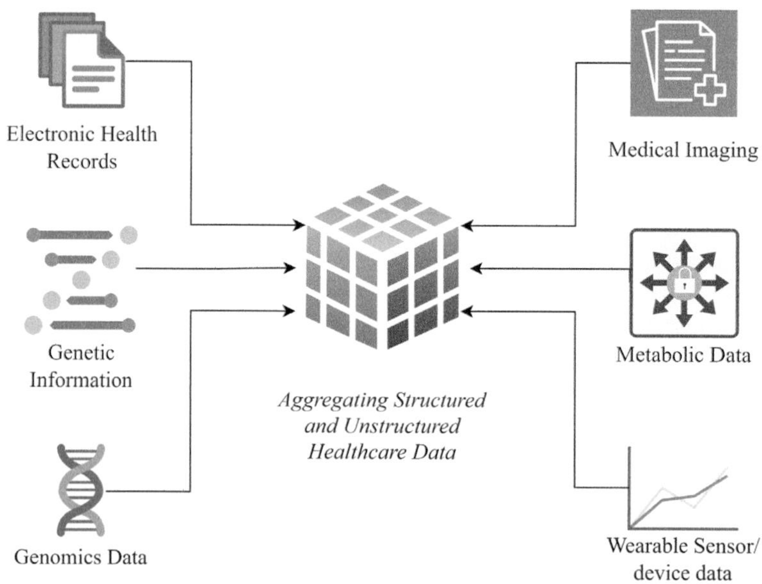

Figure 1. Heterogeneous healthcare data for disease prediction.

create intelligent models for enhanced disease prediction. Precise treatment plans and medication recommendations can be gained with data analysis. The collection and organization of data have been designed as effective storage systems to process and combine data from various sources. Data aggregation can be achieved through synchronization, where the input's diverse data sources are transformed into a standard format and stored for analysis and interpretation. Data warehouse creation is followed by feature extraction and dimensionality reduction techniques to simplify complex datasets (Waring et al. 2020). Application of advanced ML and DL algorithms has improved the efficiency to combine multimodal data, including the data inputs from textual and imaging sources to create comprehensive patient profiles (Rudrapatna and Butte 2020). Tools like Apache Spark and Hadoop are used to handle large-scale data integration, enabling real-time analysis of diverse data streams in a parallelized fashion (Ong et al. 2016).

2.2 Challenges in Integrating Heterogeneous Data Sources

There are several challenges when working with different data types, the main challenge is in integrating the data from these diverse data sources. The challenges of data integration include the variations in data formats, quality, and context. The data dimension and the type of imaging device creates issues with integration without data loss and misinformation. The Integration of structured and unstructured data into a common dimension also introduces significant challenges. Especially with the unstructured data gathering from the free-text clinical notes and medical imaging and clinical test reports. These diverse formats and their significance make it important to record and study these documents for precise treatment plans. The AI tool that is to be prepared to assist the doctors with the treatment plan should also be able to analyze these diverse inputs and give suggestions. The heavy dependence makes the systems complex and provides interoperability between the structured data from EHRs and the unstructured data elements. Additional challenges that affect the dependability of aggregated datasets include the differences in data quality, missing values, incorrect labeling, and temporal gaps (Putrama et al. 2024).

Along with the need and issues with interoperability and dimensions of the data, alignment and mapping them with alignment to multimodal data have various concerns to be addressed. Temporal inconsistencies can arise when clinical events captured in textual records do not match the time-stamped entries in structured datasets, making it challenging to create a unified timeline for patient history (Klein et al. 2000). Data privacy and security is another important factor to be considered if the personal and sensitive data of the person is being shared across these data warehouses. The data that is stored needs to be tested, and AI algorithms have to learn from these data to enhance their capacities due to issues with verification and security constraints. The sensitive patient information from different sources can expose the personal data in data breaches, if not handled with robust encryption and compliance mechanisms with the rise of risks with digital systems (Koleck et al. 2019).

2.3 Role of Data Interoperability in Multimodal Systems

Data sharing and communication should be a common technique in multimodal systems to ensure that the data are aligned into the common dimension and the inconsistencies in the data sources and timeframes can be adjusted and irregularities can be reduced. Data integration requires effective communication between diverse data sources for data interoperability. Data from one hospital sometimes need to be shared with others for treatment continuation and thus frameworks should contain a centralized repository where medical data is stored and processed. An ecosystem of medical data and patient health information has to be designed based on protocols and frameworks for data integration, operations and analysis in a robust and intelligent healthcare system. There are various challenges associated with integrating data from different institutions in the current context due to the lack of a centralized and common framework that supports data efficiency and modelling requirements (Feldman et al. 2018). By implementing interoperable standards like openEHR and HL7 Fast Healthcare Interoperability Resources (FHIR), data interchange may be significantly enhanced, and patient information can be shared easily between various healthcare organizations for the creation of intelligent healthcare systems that learns from the patterns that can be observed from the treatment histories and patient health records.

Improving data interoperability enhances the possibility of integration and supports the development of accurate disease prediction models. For example, genomic data is integrated with clinical and lifestyle data in precision medicine (Batko and Ślęzak 2022). Ensuring data consistency and including data from all the possible sources is a crucial concern for the design of interoperable systems. Real-time applications can give accurate and dynamic information obtained with the help of various sensors in real-time with a lot of information that is crucial for decision-making. Access and monitoring of precise real-time parameters of patients are particularly important in critical care applications to enhance the patient care and outcomes.

3. Application of Multimodal Systems in Decision Prediction

Multimodal AI applications with pipelines to handle the increasing complexity and heterogeneity of healthcare data have initiated the development of systems that can help adopt multimodal systems in disease prediction and patient care. Multimodal multidimensional systems integrate data from diverse sources. The systems will be able to effectively handle the diverse data formats including clinical records, genomic profiles, medical imaging, and real-time physiological signals. Intelligent medical applications can be generated setting these intelligent data processing and analyzing models as base. These systems facilitate the interaction and analysis of data from several data modalities. The information gained from the various data points provides medical conditions and attention that are required to be taken on patient health. The application of data from various sources improves the precision of treatment methods. The analysis of patient health from various data points imports the trustworthiness and completeness of AI models with respect to the data that is used to make decisions.

Various applications using multimodal data fusions are designed with the aim of data collection and design of AI systems; each of the systems has focused on data storage and integrations strategies. A complex system with the combination of these individual systems has to be designed to improve the efficiency in data modelling for healthcare applications. There are different data processing strategies and applications designed to handle the multimodal data. The application of these intelligent systems on diagnosis and prognosis are highlighted and their potential to enhance data driven clinical decisions made to improve patient outcomes are explained. The ability to include data from other sources including genetic information, imaging data and clinical observations are integrated into the precision medicine systems. There is a huge potential for these intelligent tools to help develop personalized treatment plans and facilitate early illness identification.

3.1 Case Studies of Multimodal Data Fusion for Disease Diagnosis

Multimodal data fusion has been applied in various clinical contexts to improve diagnostic accuracy. Clinical studies have integrated different brain imaging modalities, such as functional Magnetic Resonance Imaging (fMRI) and PET, for improvements in diagnosing different neurodegenerative diseases like Alzheimer's (Das et al. 2021). The network fusion approach combining data from genomics and neuroimaging was done to find biomarkers to identify the early illness stages (Wang et al. 2014). The AI and data analysis solutions made it possible to recognize and quickly identify complex relationships in patient conditions, giving a more comprehensive depiction and treatment support for various diseases. The data analysis used data from different sources and processed them to gain insights. Complex patterns were mined from these sources with the help of AI algorithms.

To improve the identification of neurological illnesses, researchers used optimized biologically inspired neural networks to perform multimodal image fusion, merging CT and MRI data (Daniel et al. 2017). Compared to single-modality models, the fusion models could significantly increase the accuracy of lesion identification. Multimodal DL models have been implemented to predict cancer subtypes by integrating histopathological images and multi-omics data providing more precise disease characterization. Application of intelligent solutions and to speed up the clinical trials and treatment plan preparation has been done with the support of ML and DL algorithms and its capacities. These algorithms with the power to analyze data and gain insights can be applied in medical applications to enhance treatment plans and comprehensive models in processing and analyzing data.

3.2 Combining Genomic, Imaging, and Clinical Data for Disease Prediction

Understanding complicated illnesses impacted by numerous biological variables is easier by combining genetic, imaging, and clinical data. Figure 2 represents the different data dimensions that can be integrated as part of a comprehensive health

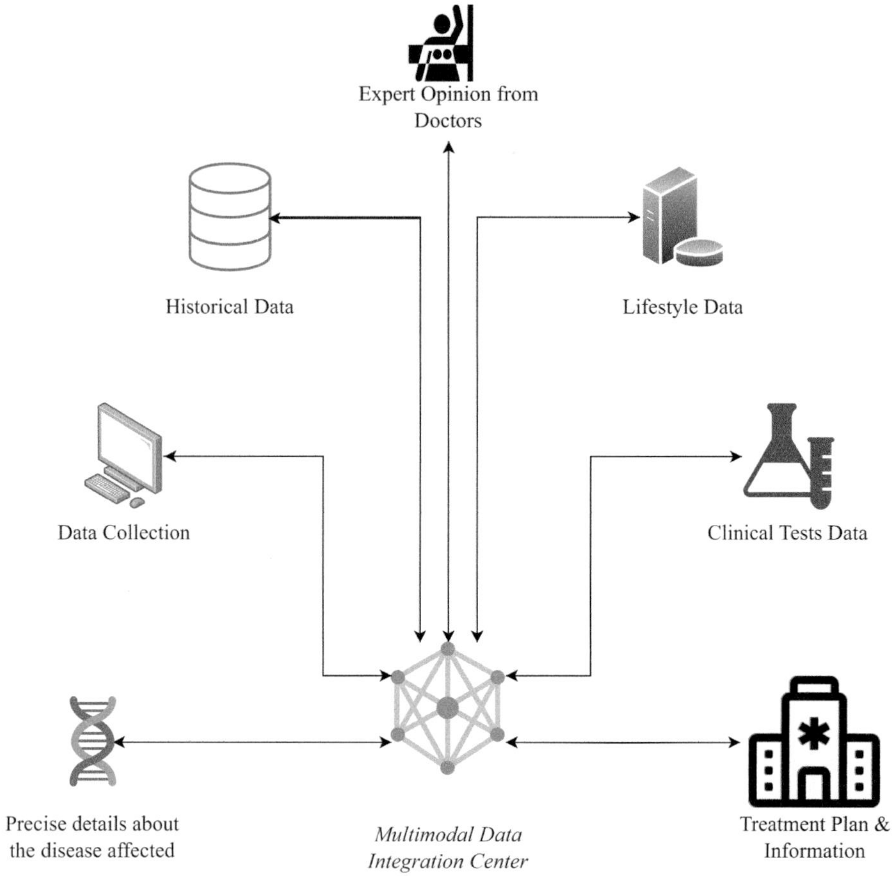

Figure 2. Process of precision medicine using multimodal data.

plan. The data includes imaging data that can give information about the body parts, structural and functional changes in organs and tissues. Genomic data gives information on the hereditary traits and characteristics. Lab reports that give an insight to patient histories are few of the data that can be integrated for improving medical findings. A multimodal data integration center, that is part of the AI application to analyze data and give predictions, helps in processing and aggregating data. Precision medicine system is made possible by combining different data sources to create prediction models that account for genetic and phenotypic differences (Bao et al. 2022). Joint sparse canonical correlation analysis is one of the methods that was used to integrate genomic and imaging data. This approach could identify neuro markers linked to schizophrenia, offering insights into the interaction between genetic variations and abnormalities in brain structure (Li et al. 2009). Multimodal systems capable of working with multiple data sources that can combine Single Nucleotide Polymorphism (SNP) data with MRI scans have been used to predict the risk of developing neurological conditions, allowing for early intervention and management (Sui et al. 2013). There are various applications of case-to-case aggregation of data from various sources for disease prediction and treatment plan preparation.

3.3 Role of Multimodal Data in Precision Medicine

Precision medicine is the process of preparing customized treatment plans for patients. Gathering patient data from various sources can achieve this. This includes data from genetic, environmental, and lifestyle dimensions. The precision treatment approach helps prepare specialized treatment strategies for the patient. This needs data analysis from several sources to provide various patient health issues. In precision oncology, for example, integrating genomic data with histopathological images has enabled the development of predictive models that can classify cancer subtypes and recommend targeted therapies (Qi et al. 2017). These intelligent AI models that analyze patient data have been designed on a smaller scale. The main challenge in the creation of generalized models is data availability. Along with data availability, issues include the computational complexity and the lack of centralized agencies to monitor the design and development of such intelligent models. Multimodal systems have been applied in rare disease research, where data from different modalities are integrated to overcome the limitations of small sample sizes, typically seen in single-modality studies (Lahat et al. 2015). Combining data from multiple sources for analysis has helped understand the genetic basis of rare diseases in patients and helped prepare personalized treatment plans. The understanding and processing of genomic data requires complex planning and analysis pipelines. In most cases, a general pipeline will process and prepare the data for subsequent independent analysis, based on the outcome that is expected from the data processing and analysis.

4. Machine Learning and Deep Learning Models for Multimodal Data

Integrating multimodal data in healthcare is becoming complex because the different types of data sources and their representations make it challenging to apply intelligent algorithms directly into the AI models, for disease prediction and treatment planning. ML and DL algorithms can create models from a single modality by directly applying them to data. Using the feature extraction and multimodal approach, AI models can handle diverse data sources for creating intelligent models. These hybrid approaches can handle data from wearable sensors, genomic data, clinical records, and imaging modalities. To enable complex analyses that are impossible with single-modality, multimodal ML algorithms can effectively recognize the complex relationships from the different data that can be related. Table 2 compares the different ML models for the multimodal disease prediction application. To ensure that all data modalities are included and effective modelling using various data modalities, hybrid approaches with fusioning and voting mechanisms are to be designed to work the various ML and DL algorithms to improve the prediction accuracy.

Table 2. Comparison of machine learning models for multimodal disease prediction.

Model Type	Description	Applications in Healthcare	Key Advantages	Limitations
Multimodal Autoencoders	Models that compress input data from various modalities into a shared latent space and reconstruct it.	MRI data integration, Genomic data imputation	Effective in handling missing modalities	Sensitive to noise in data
Hybrid Deep Learning Models	Combines deep neural networks with classical ML algorithms (e.g., CNN + SVM).	Predicting diabetic retinopathy, Cancer diagnosis	Improved classification accuracy	Computationally intensive
Attention-based Models	It uses attention mechanisms to focus on relevant features from each modality.	Disease progression tracking using EHR and imaging	Captures complex inter-modality relationships	Requires large training datasets
Multi-task Learning (MTL)	Simultaneously learns multiple related tasks, sharing representations between them.	Predicting multiple disease subtypes, Drug response prediction	Reduces overfitting, effective for small datasets	Task interference can reduce performance
Transfer Learning	Uses pre-trained models on large datasets and fine-tunes them for new tasks.	Genomic and imaging data analysis	Reduces training time, works well with limited data	Risk of negative transfer if source and target tasks differ significantly
Generative Adversarial Networks (GANs)	It uses generator and discriminator networks to generate synthetic data.	Creating synthetic multimodal data for rare diseases	Addresses data imbalance	Unstable training.

4.1 Multimodal Machine Learning Architectures in Healthcare

ML algorithms are designed to handle multimodal data to effectively process healthcare data. Multimodal ML algorithms have the ability to integrate and process information from multiple sources. A typical multimodal ML architecture involves three key components:

- Modality-specific feature extraction
- Shared representation learning
- Decision fusion.

The feature extraction phase is a set of pipelines that are enabled to process data to extract meaningful data. A separate preprocessing pipeline is required to extract data from each of the data types, Convolutional Neural Networks (CNNs) for imaging data and transformers for textual data, to obtain specialized features from each modality (Zhang et al. 2020). The extracted features from the preprocessing pipelines are then mapped into a shared latent space using joint or coordinated representation strategies to capture cross-modality interactions and correlations (Srivastava and Salakhutdinov 2012). The preprocessing pipeline contains ML models and data cleaning steps to clean and organize the data into a usable format. A shared representation learning environment needs to be designed with the data that is preprocessed and organized. ML and DL algorithms have been trained, tested and retrained using data to improve the efficiency of the models. The decision fusion layer helps to take the best results from the ML models using a voting mechanism, where the best results from the models are identified and selected as the prediction result.

In multimodal ML architectures, using attention mechanisms to selectively focus on relevant features from each modality, is one of the important factors that is to be considered while designing feature extraction framework. For example, an attention-based fusion technique has been applied to integrate EHR data and radiological images. This significantly improved the prediction of disease progression in patients with chronic illnesses (Kumar et al. 2023). Graph-based learning models represent multimodal data as interconnected nodes, and have also emerged as a promising technique for disease prediction, especially in complex cases involving interactions between genetic and phenotypic factors (Wang et al. 2014). Different models and frameworks have been designed and developed by researchers; these models and frameworks have to be studied to apply them on the real-time data. These models and frameworks work better with the data that they have been trained with and exposed to. Real-time application of these models requires more comprehensive evaluation techniques and constant monitoring of the results have to be employed and the predictions have to be in a supervised mode.

4.2 Deep Learning Approaches for Multimodal Data Fusion

DL (Deep Learning) methods have made a significant contribution to multimodal data fusion. With the help of advanced architectures, such as multimodal autoencoders, deep generative models, and adversarial networks, significant learning can be done from the data. Multimodal autoencoders extend the traditional autoencoder framework by encoding information from multiple modalities into a shared latent space, which can then be decoded to reconstruct any missing or corrupted modality (Zhang et al. 2020). The application of multimodal autoencoders has been efficient in scenarios, where one modality of data is missing or incomplete. The processing is done with imputation based on complementary modalities (Kumar et al. 2023). Variational Autoencoders (VAEs) and Generative Adversarial Networks (GANs) are used for modeling complex dependencies across modalities.

The application of DL approaches for data processing and modelling in medical data requires huge consideration. The working of models and the features learned from the data cannot be described and the black box nature of these models creates challenges in understanding and explaining model classification. Explainable architectures are designed but need to get more compatible with an increasing number of DL algorithms. Using these AI algorithms, researchers also create synthetic data to address the issues with data and focus on model development. The model created using VAE and GAN with synthetic genomic and imaging data could improve prediction results (Bao et al. 2022). The lack of proper ML algorithm training and learning creates issues with data alignment between modalities. The unbalanced data affects the model performance of the AI (Lahat et al. 2015). When not normalized and when it is from relative sources, the data creates an imbalanced dataset for AI algorithms that tend to create bias. Data issues can be fixed by gathering reliable data to develop intelligent models under regulatory frameworks.

4.3 Hybrid Models - Combining AI with Conventional Techniques

Hybrid models are designed to address issues that traditional models fail to handle with data from diverse sources and associated complexities. The hybrid models combine DL and ML techniques to use their advantages and ensure the data is appropriately managed, and predictions happen correctly. For example, hybrid frameworks could incorporate ensemble approaches for decision-making or traditional regression models with deep neural networks for feature extraction. The layered approach performs better with the complex medical scenarios, where multi-disease risk prediction, and therapy optimization using personalized treatment approaches, enable a more detailed understanding of the multimodal interactions of components in the data (Qi et al. 2017).

Figure 3 represents a hybrid ML architecture for multimodal data processing using healthcare data. The hybrid model, with the combination of CNN based feature

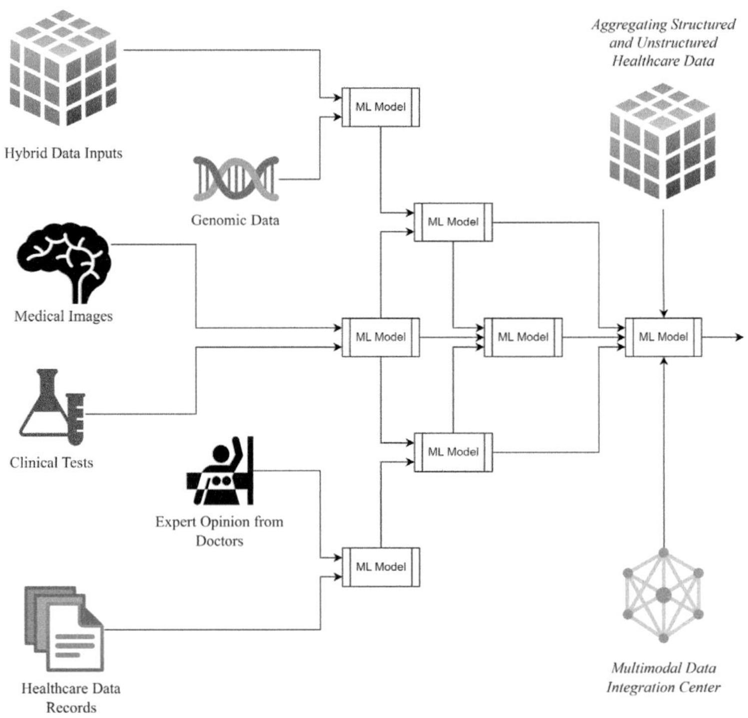

Figure 3. Hybrid machine learning architecture for multimodal healthcare data.

extraction from images and Support Vector Machine (SVM) for classification, has been used to predict the presence of diabetic retinopathy using retinal images (Duan et al. 2020). To improve diagnostic accuracy in the uncommon disorders, data from various sources plays a huge role. Medical imaging (CT, MRI) and EHR have been integrated using hybrid DL frameworks (Lahat et al. 2015). The interpretability of the results and working of the classical ML models, and the feature extraction and learning capabilities of DL networks, are combined to create hybrid models. The hybrid models offer a reliable way of handling and interpreting complex multimodal healthcare data.

5. Intelligent Algorithms for Multimodal Disease Detection

Integrating intelligent AI algorithms for the multimodal systems has the potential to revolutionize the way in which disease identification has been happening. Intelligent integration of data sources, includes the integration of medical imaging, clinical records, and genomic information. Data integration and processing makes the analysis more complex. The application of AI algorithms is used for identifying and extracting relevant features. The model fine tuning process helps to select the best models and parameters to capture complex interactions in the data. Various models applied for multimodal disease detection work with effective feature extraction and parameter selection techniques. The intelligent approach focuses on multi-task model selection and transfer learning models. Various evaluation metrics help to assess the performance of the individual models.

5.1 Algorithms for Feature Extraction and Selection in Multimodal Systems

Feature extraction and selection are important steps to be taken with data, especially with multimodal data integration and processing. This process of feature extraction and selection helps with the reduction of dimension, minimizes noise, and enhances the predictive power of the data and the AI model, that is designed using the fine-tuned data. The main challenge with the multimodal systems is keeping the feature information from each of the modalities and processing it to combine them. Each data in its form has varying characteristics which makes these modalities important. The processing of some of the data such as clinical writings made by the doctors and nurses about the patients sometimes have observations that are obsolete. The medical imaging data that is captured sometimes may not carry the required information in one modality, making it a necessity to depend on another modality. To capture spatial and structural characteristics from the imaging data, DL approaches like CNNs are combined with traditional feature extraction techniques including Principal Component Analysis (PCA) and Independent Component Analysis (ICA) (Bhutto et al. 2024).

For example, Wavelet transform and grey-level co-occurrence matrices are used to extract textural characteristics from radiological pictures. The features extracted with Wavelet transform and grey-level co-occurrence matrices are subsequently merged with clinical data from EHR using a hybrid algorithm that uses fuzzy entropy for feature selection (Tawfik et al. 2022). Significant characteristics have been identified and selected from different modalities using dictionary learning and sparse representation, enabling the creation of strong representations that improve the sensitivity and specificity of illness detection models (Liu et al. 2016). Complex algorithms and pipelines have been proposed by various researchers to process and integrate multidimensional data.

5.2 Multi-task Learning and Transfer Learning for Disease Prediction

Multi-task learning (MTL) and transfer learning are effective methods in multimodal illness diagnosis. These methods can share the information for model development and can store the patterns identified from the dataset for the model fine tuning process. The techniques MTL and transfer learning are effective methods in multimodal illness diagnosis. MTL reduces the danger of overfitting of data in models by utilizing similarities to enhance the performance of specific tasks, when labeled data is limited. Utilizing common representations from genomic and imaging data, MTL may concurrently predict various outcomes (e.g., distinct cancer subtypes) in disease prediction models (Lahat et al. 2015).

Transfer learning helps to apply the weights and features, that have been learned from a previously trained model on a related task, to a new but related problem. The multimodal systems benefit from transfer learning methods since not all modalities have massive datasets for ML models to learn. A common approach in building complex models that excels in a problem is to use a pre-trained model that has been trained on a large dataset and fine-tune the model using the specific data that is related for the problem, then evaluate the performance of the model and then train it again to make it more effective. This approach is used for the processing of medical imaging tasks, such as MRI or CT scan analysis (Zhang et al. 2020). Transfer learning has also been used to integrate genomic and imaging data by using pre-trained networks to extract features from data and combining them with genomic embedding. The transfer learning approach results in improved predictive performance for complex diseases and has been applied for diseases like Alzheimer's and Parkinson's (Zhang et al. 2020).

5.3 Evaluation Metrics for Multimodal Models

The performance of AI models needs to be evaluated and monitored when applied for decision making. These methods help to measure the accuracy and efficiency of the models. The performance of AI systems has to be measured and the results from these systems have to be cross checked for the removal of potential data loss and misinformation. An evaluation process is required when the data is processed using fusion approaches and multimodal data is used by the algorithms to learn. The efficiency of the fusion techniques has to be evaluated to ensure proper alignment of the data and model learning capability. Standard evaluation measures are accuracy, sensitivity, and specificity. The use of standard metrics is not fully able to evaluate the complexity of multimodal data. More measures need to be applied to assess the performance of AI systems and ML algorithms designed to work with the medical data (Chen et al. 2024).

In addition to the traditional evaluation metrics (accuracy, sensitivity, and specificity), cross-validation and ensemble techniques need to be used to validate the robustness and performance of multimodal models. The working of models also needs to be compared with different datasets. Cross-validation ensures that the model generalizes well by testing it on multiple data splits. The ensemble methods such as bagging and boosting, combine the predictions from various models to improve stability and accuracy (Dal et al. 2023). Along with the evaluation metrics, interpretability of the measures and models have to be tested. Performance evaluation using the measures—Shapley Additive Explanations (SHAP) and Local Interpretable Model-agnostic Explanations (LIME) are applied to understand the contribution from each of the modalities to the final prediction in a multimodal system, where AI models are involved. The SHAP and LIME methods help to understand the contribution from each of the data modalities in the decision-making process in a multimodal system (Wang et al. 2024). Different evaluation techniques (traditional and customized techniques) are applied to evaluate the performance of the AI models to improve the performance and application space of the models.

6. Precision Medicine and Multimodal Data Processing

Precision medicine helps to improve the accuracy and efficacy of disease diagnosis. With the help of data, targeted therapy (precision medicine/personalized treatment) gives better patient care and improves the traditional treatment with specific patient care. A fundamental component of this strategy is integrating data from data types, including behavioral, lifestyle, and genetic data. Integrating several data sources provides a better understanding of patient health, which is impossible with conventional systems. Multimodal data integration, personalized treatments, and outlines of future directions need to be analyzed, and practical frameworks for integrating and creating a complex database are crucial for healthcare systems. AI and ML algorithms help process this extensive data from various sources. The AI application needs real-time analysis and targeted personal care interventions.

6.1 The Role of Precision Medicine in Improving Prediction Accuracy

Precision medicine aims to provide personalized treatment for patients. In precision medicine, data analysis plays a huge role in identifying patient issues. Special attention is given to data obtained from various sources including the variations in the genomic structure, structural and functional composition and its various, real-time vitals. Using the information associated with the patient at various levels, AI models can be used for monitoring and identifying hidden patterns in the data. AI models and predictions help plan for the preparation of precise treatment plans with higher accuracy. These models have the capacity to combine data from multiple sources that give insights about the patient's current conditions. Similarity network fusion is a technique that is used to aggregate data from different data types at the genomic level. The similarity network fusion technique has been effective in identifying biomarkers that are the cause for disease progression (Wang et al. 2014). Integrating data from diverse sources helps to uncover complex patterns of interaction between various factors affecting diseases. Complex algorithms help in the study and identification of disease and its subtypes. This helps in the development of precise diagnostic tools (Bao et al. 2022). These intelligent techniques help in the analysis of data and preparation of precise treatment plans.

The issues with applying AI and intelligent techniques for the preparation of precise treatment plans and analysis of the data needs a comprehensive system. This system can process data from various sources and standardize the data and record them in a centralized data warehouse. When analyzing patient data, the data should be secure and the system should ensure that the data is not compromised. The data analysis helps in early identification and provides interventions for illnesses, including cancer and cardiovascular problems (Chen et al. 2024). Precision medicine has huge potential for the development of intelligent patient care systems. The application of AI and advanced predictive systems will empower the future of healthcare systems that focus on patient care and provide targeted interventions when required.

6.2 Personalization of Treatment Through Multimodal Data Integration

Personalized precise treatment plan preparation needs the understanding of the patient and his genomic and genetic information. The data from sources like genomic information, and other patient characteristics from multiple data sources, is crucial for making informed decisions. This information is then mapped with various databases that store the information about the conditions and drugs that will work for specific conditions. The information will give the insights on the possible medications that will work effectively for the patient. Using advanced multimodal DL algorithms, this analysis on the genomic data, medical imaging and clinical history along with other sources gives an overview to the disease and the possible treatment plans. The analysis of disease progression helps to prepare personalized treatment strategies based on individual profiles (Zhang et al. 2020). The multimodal data analysis frameworks powered by AI helps to map and extract information from various modalities. The application of these intelligent multimodal data processing frameworks show improvement in the recommendations of treatment plans for diseases like diabetes and neurological disorders (Wang et al. 2024). For the treatment plan preparation of cancer patients, the analyzing of the histopathological data with multi-omics data (proteomics and metabolomics) has provided personalized treatment plan recommendations. The recommendation of personalized medicines that reduce side effects and enhance patient outcomes has been made possible by the preparation of personalized treatment plans with successful identification of patient-specific pathways that support tumor growth. The detailed analysis of the data from various points helps to design personalized treatment plans for the patients to help them with various disease conditions.

6.3 Future Directions: Combining Genomic, Lifestyle, and Behavioral Data

The design of multimodal data processing and predictive AI models must be able to handle data from different sources. These data sources include processing of patient behavioral factors including lifestyle data, clinical records, and genomic profiles. This data-driven treatment approach is the key to the future of precision medicine. Effective data processing and integration frameworks have to be designed to make data integration effective. Real-time lifestyle data via wearable sensors and behavioral records helps to identify the hidden patterns that cause the disease progression and analysis. The future of precision medicine includes gathering patient genetic composition and understanding the changes caused in the genetic structure by the disease. The process of combining data from genomic modality with others will help to identify the cause of a disease. Identification of the cause of the disease helps in the preparation of treatment plan for the patients (Dal et al. 2023). This process of analyzing data and gathering information about the patient health gives the recommendations for personalized treatment based on each patient lifestyle. The precise treatment process accounts for the dynamic changes that the patient will have during the treatment process. Multimodal systems can analyze the data and forecast chronic diseases and their outcomes based on the genomic data and physical activity levels, sleep patterns, and dietary habits. Multimodal models are designed to detect and give early warning indicators of disease and suggest preventative steps by examining the relationships in the data (Shaban-Nejad et al. 2022). Active health management is achieved with the integration of AI data processing models, which helps in monitoring an individual and providing treatment plans that need to be regularly updated and adapt to the constant changes that have been caused. The systems are working based on real-time patient data and set the future of treatment planning and patient care.

7. Potential and Future Trends in Multimodal Disease Prediction

Detection of diseases from the multimodal data source related to the individual is made possible with the capture of various data points at its analysis. The preparation of appropriate treatment plans based on the recommendations from the data that is captured for the analysis is gaining significance in healthcare. The multimodal system uses the advancements in AI, ML, and data fusion techniques. The system captures and integrates diverse data modalities including the data from medical imaging, genomics, and patient mentioned conditions. The multimodal systems can enhance predictive accuracy and facilitate early disease detection. There are various emerging technologies in the multimodal healthcare systems. These systems are aimed to improve the outcomes of each of the applications in their users. There's a huge potential for the multimodal approach of intelligent healthcare systems in telemedicine and remote monitoring, and the ethical considerations shaping the future of multimodal data processing in clinical practice.

7.1 Challenges in Multimodal Data Processing

The multimodal data processing approach for healthcare has huge possibilities. However, there are several challenges associated with the application of these intelligent systems. The challenges of these systems must be addressed before the integration and powering the systems to manage diverse data sources. The main difficulty is the lack of data or the missing parameters in the data. Along with these missing or issues with data structure, their data privacy and security issues are also to be considered and addressed. The number of computational requirements and the data size are some other challenges. Bringing standards in data processing is one of the approaches towards resolving the problems with scalability, robustness, and dependability of multimodal systems. These standards help design robust and powerful systems with support for automation and wide integration options. Providing options to address the issues with multimodal data processing.

7.1.1 Data Privacy and Security Concerns in Multimodal Systems

Building intelligent systems needs large-scale collection and curation of data. This data is often related to a person. Handling the data, to build intelligent systems to support efficient and precise patient care, without it being misused is crucial. Ensuring privacy and secured access to sensitive patient data is one of the main issues that must be addressed. The data from diverse sources (medical imaging data, clinical records, genomic data, and so on), make it challenging to ensure compliance with privacy laws like HIPAA. This complexity increases as there is no centralized agency with different intermediaries to coordinate and monitor when data is shared across multiple institutions and processed. The risk of unauthorized access and data breaches of the data in data warehouses increases the privacy breach challenges and the data breach leads to data misuse. Advanced encryption methods, federated learning, and privacy-preserving algorithms enable the secure access of the data from the data warehouse for analysis and modeling. AI models cannot be built without direct access to data. Direct access to personal and sensitive data should be monitored by a central authority. The system needs to effectively handle diverse data with security and privacy and open the door for innovations in healthcare systems (Dal et al. 2023). These policies and frameworks, designed to protect and regulate data access, can reduce security threats and facilitate cooperative research by safeguarding patient privacy.

7.1.2 Handling Missing and Incomplete Data in Healthcare Applications

Multimodal healthcare systems often suffer from the issue of data incompleteness. The issue of data incompleteness is due to various factors; it can be because of issues during the time of data capture, missing records, lack of access, or type mismatch. The incomplete data when used for the modelling or testing process can affect the performance of the AI models. The lack of data makes the system blind in certain areas, and this makes it challenging to address the complexities associated with that area or dimension. The issues with the data, lead to biased results or reduced accuracy. Data imputation methods have been used as a solution to address the challenge with respect to missing data. Use of generative models create similar data to fill in the missing data. Variational Autoencoders (VAEs) is a generative technique that is used to estimate the missing values based on understanding the patterns and relationships from the available data. Using VAEs enhances the robustness and reliability of ML models in multimodal healthcare systems. VAEs ensure that the models can work better with the data by filling the missing values with the most relatable data based on the available information (Chen et al. 2024). Multi-view learning approaches can effectively handle scenarios with missing modalities and data. The missing data modalities have been told to models in the multi-view learning approach to ensure that models can function even when some data sources are unavailable (Zhang et al. 2020). Significant improvements in bringing standardization are required to bridge the issues with missing values in various dimensions.

7.1.3 Computational Challenges and Resource Management

Processing multimodal data in real-time has high computational requirements. The need for high computational requirements is due to diverse data formats and the scale of dimensions in some of the data modalities. The training of multimodal AI systems requires large-scale storage, processing, and complex training requirements. For training advanced AI models, there is a need for the use of distributed computing environments and advanced hardware. The advanced hardware includes Graphics Processing Units (GPUs) and Tensor Processing Units (TPUs) (Chen et al. 2024). Implementing or applying these heavy multimodal systems in clinical settings also requires heavy computational resource allocation and management. Distributed learning, model compression, and parallel processing can handle these challenges of heavy computation requirements. These solutions help to guarantee that multimodal systems can grow effectively without sacrificing performance (Wang et al. 2024). The development of these intelligent models requires access to heavy computation requirements. Effective and accurate models need to process large chunks of data and understand the patterns from them. This emphasizes the need for large scale computation requirements for training and testing the models.

7.2 Emerging Technologies in Multimodal Healthcare Systems

Emerging technologies in multimodal healthcare systems are revolutionizing healthcare data collection, integration, and analysis. Advanced ML frameworks such as DL transformers and graph-based models are recent inventions that provide enhanced capabilities for managing intricate multimodal datasets (Shaban-Nejad et al. 2022). Advanced AI models can forecast diseases by identifying the complex interactions between data and various modalities. Non-invasive imaging and medical techniques along with more advanced wearable devices are becoming popular. These devices can provide data that have large storage requirements. AI-enabled wearable technology makes it possible to constantly monitor the patient and analyze the data in real-time. Real time data analysis makes it useful for controlling various illnesses. To facilitate real-time health monitoring, sensors are used to collect physiological signals and send them to cloud-based platforms to integrate clinical and genetic data (Dal et al. 2023, Liu et al. 2023). AI-enabled methods, like multi-scale picture fusion and 3D image reconstruction, are enabling medical imaging to combine several imaging modalities, including PET, CT, and MRI scans, to improve diagnosis accuracy (Wang et al. 2014). There are a lot of innovations and solutions that have been designed and developed. There is a need for a centralized system that can aggregate these individual components to act as a virtual assistant.

7.3 Use of Multimodal Approaches in Telemedicine and Remote Monitoring

Patient care system has been transformed and innovations are common in healthcare applications. The use of multimodal telemedicine applications and remote monitoring techniques have been designed to assist the patients from home without having to physically come to hospitals. These systems can also be beneficial for people with limited access to healthcare facilities. There is a need for the integration of the data from wearable technology, EHRs, and live video consultations for the development of the virtual telemedicine ecosystem. This integration makes continuous health monitoring, early intervention, and individualized treatment plans based on multimodal data inputs possible (Chen et al. 2024). The patient's vital indicators (blood pressure, heart rate, oxygen saturation, etc.) can be tracked using wearable equipment. These wearable gadgets send data to the data analysis stations and to the medical professionals (in hospitals), which are then analyzed using AI algorithms to identify abnormalities and evaluate trends in the data. Telemedicine is becoming popular and effective as on-the-spot evaluations are provided using remote diagnostic equipment. The AI-enabled portable ultrasound devices (Shaban-Nejad et al. 2022) help with easy ultrasound scans on patients, and is an example of easy to use, hand held devices. Healthcare practitioners can comprehensively understand a patient's health state by integrating telemedicine equipment with cloud-based EHRs. This arrangement makes better decision-making and fewer in-person visits possible by increasing access to healthcare and enhancing patient outcomes (Lin et al. 2014).

7.4 Advances in Real-time Multimodal Data Processing for Predictive Analytics

Real-time multimodal data processing helps in the significant predictive analytics within healthcare. Real-time data analysis helps in dynamic patient monitoring and timely intervention. Combining data from wearable devices, EHRs, and various other sensors enables healthcare systems to develop predictive models capable of identifying health risks and forecasting clinical events, including disease exacerbations, before clinical manifestation. To manage high-dimensional data streams, real-time data processing systems in the healthcare industry use machine learning algorithms and other prediction approaches. These systems combine patient history and continuous physiological parameters to generate comprehensive, context-aware health profiles. With the system getting complex, addressing the issues with computing needs and data interoperability is crucial to enhancing the scalability. Data processing for multimodal systems is heavy. The data from diverse forms are to be combined and analyzed for information. Smoothly integrating data from different sources requires policy frameworks and shared planning and standards.

7.5 Ethical Considerations and the Future of Multimodal Data Processing

Development in AI and healthcare systems is rapid. Constant updates are present in the various components of AI. Effective architectures are designed for multimodal data processing in healthcare. There are ethical issues connected with the development of these AI models. These issues are due to computational limitations, algorithmic bias, data privacy, and informed consent. Trusting the AI models for analyzing and processing the data by protecting the patient data with strong security protocols is to be considered. The model development also should ensure that the AI models reduce the biases in the data and should not make wrong predictions. To guarantee the fair and ethical use of the intelligent AI systems, dynamic identification and addressing issues are required (Chen et al. 2024). A framework also has to be designed for multimodal systems that can address the data security, privacy and operational challenges. To guarantee that the predictions produced by multimodal models are understandable and explainable, the design of explainable AI (XAI) models are required. The major requirement for the intelligent applications is to explain the findings, this can be achieved with the help of XAI. In order to protect the patients personally identifiable information (PII), there is a need for creating strong data governance standards. These are specially needed with these systems for sharing data among multiple organizations. These guidelines guarantee adherence to privacy norms and encourage the safe, moral use of data in healthcare technology development (Wang et al. 2024). Future directions in multimodal data processing will likely involve greater real-time behavioral and environmental data integration. The data integration and processing approach need federated learning and multi-party computation to preserve patient privacy (Dal et al. 2023). Innovation-driven solutions must get an ethical balance and protect individual rights. Achieving the balance between innovation and privacy is to be given top priority.

8. Conclusion

Data is crucial in providing insights on the medical conditions in healthcare systems. The data gathered from various sources helps to make informed decisions with the support of data analyses. Multimodal data processing systems are required to collect, combine, and analyze these data to gain insights from the healthcare data. These healthcare systems help in disease detection, prediction, and patient management. Integrating diverse data sources, including clinical records and imaging, and genomic and behavioral information will significantly enhance the accuracy and robustness of predictive models made with the integrated data sources. Multimodal systems have improved and applied complex ML and DL algorithms to process and combine patient data. Multimodal data processing systems in healthcare are one of the preliminary requirements for extensive precision medicine. Precision medicine helps in individualized treatment planning and illness diagnosis and significantly assists patient care.

Multimodal systems are designed to gather and combine data from diverse sources. These systems produce a centralized data warehouse to produce precise medical decisions. The AI models that make use of the data from these complex processing help to enhance risk assessment and disease predictions. These intelligent systems understand the patient and the disease with the help of data processing and analysis. The systems can combine and organize various structured and unstructured data. The data from various sources helps the intelligent model to capture connections between various data modalities. The application of multimodal DL models, such as multimodal auto-encoders and hybrid architectures, has shown promise in improving prediction accuracy for complex diseases, such as cancer and neurological disorders. Medical professionals may use various data sources and insights from their analysis to create precise medical decisions specific to patients. Identifying the issues with the patient and recommendation of treatment plans is purely based on the concepts of precision medicine. EHRs create patient profiles and are used to document the treatment recommendations to improve the effectiveness and reduce side effects. The use of multimodal systems in telemedicine and remote monitoring has improved access to healthcare by allowing for quick treatments and ongoing health tracking for patients who might not have easy access to traditional healthcare facilities.

There are several important research issues in developing multimodal healthcare systems. The different varieties of data and the diverse formats in which these data exist are one of the primary concerns. Sharing the data to build complex systems also creates privacy and security issues, another factor affecting innovations. The future of intelligent healthcare systems will depend on privacy and data security frameworks and guidelines. The diverse data forms must be structured to ensure the data formats can be integrated in real-time. Such multimodal frameworks have to evolve by addressing the computational and algorithmic challenges. These issues will be resolved if multimodal systems are implemented on a larger scale in clinical settings. Further research and policy formulations are required to overcome the challenges. Multimodal data processing offers enormous potential to enhance patient care and disease prediction. Technological advancements help address the challenges associated with healthcare systems. More technological evolutions are required to make these systems more accurate, secure, and accessible healthcare systems that can transform patient outcomes.

References

AlSaad, Rawan, Alaa Abd-Alrazaq, Sabri Boughorbel, Arfan Ahmed, Max-Antoine Renault, Rafat Damseh et al. (2024). Multimodal large language models in health care: Applications, challenges, and future outlook. Journal of Medical Internet Research 26: e59505.

Bao, Feng, Yue Deng, Sen Wan, Susan Q. Shen, Bo Wang, Qionghai Dai et al. (2022). Integrative spatial analysis of cell morphologies and transcriptional states with MUSE. Nature Biotechnology 40(8): 1200–1209.

Batko, Kornelia and Andrzej Ślęzak. (2022). The use of big data analytics in healthcare. Journal of Big Data 9(1): 3.

Bhutto, Jameel Ahmed, Jiang Guosong, Ziaur Rahman, Muhammad Ishfaq, Zhengzheng Sun and Toufique Ahmed Soomro. (2024). Feature extraction of multimodal medical image fusion using novel deep learning and contrast enhancement method. Applied Intelligence 54(7): 5907–5930.

Chakravarthi, Bhuvaneshwari, Sin-Chun Ng, Ezilarasan, M. R. and Man-Fai Leung. (2022). EEG-based emotion recognition using hybrid CNN and LSTM classification. Frontiers in Computational Neuroscience 16: 1019776.

Chen, Xieling, Haoran Xie, Xiaohui Tao, Fu Lee Wang, Mingming Leng and Baiying Lei. (2024). Artificial intelligence and multimodal data fusion for smart healthcare: topic modeling and bibliometrics. Artificial Intelligence Review 57(4): 91.

Dal Mas, Francesca, Maurizio Massaro, Pierluigi Rippa and Giustina Secundo. (2023). The challenges of digital transformation in healthcare: An interdisciplinary literature review, framework, and future research agenda. Technovation 123: 102716.

Daniel, Ebenezer, Anitha, J., Kamaleshwaran, K. K. and Indu Rani. (2017). Optimum spectrum mask based medical image fusion using Gray Wolf Optimization. Biomedical Signal Processing and Control 34: 36–43.

Das, Manisha, Deep Gupta, Petia Radeva and Ashwini M. Bakde. (2021). Multi-scale decomposition-based CT-MR neurological image fusion using optimized bio-inspired spiking neural model with meta-heuristic optimization. International Journal of Imaging Systems and Technology 31(4): 2170–2188.

Duan, Junwei, Jiaqi Xiong, Yinghui Li and Weiping Ding. (2024). Deep learning based multimodal biomedical data fusion: An overview and comparative review. Information Fusion 102536.

Feldman, Keith, Reid A. Johnson and Nitesh V. Chawla. (2018). The state of data in healthcare: path towards standardization. Journal of Healthcare Informatics Research 2: 248–271.

Klein, Katherine J. and Steve W. J. Kozlowski. (2000). A multilevel approach to theory and research in organizations: Contextual, temporal, and emergent processes. Multilevel Theory, Research, and Methods in Organizations: Foundations, Extensions, and New Directions 3–90.

Koleck, Theresa A., Caitlin Dreisbach, Philip E. Bourne and Suzanne Bakken. (2019). Natural language processing of symptoms documented in free-text narratives of electronic health records: a systematic review. Journal of the American Medical Informatics Association 26(4): 364–379.

Kumar, Vipin, Pawan Kumar Chaurasia and Prem Shankar Singh Aydav. (2023). Deep multi-view learning for healthcare domain. pp. 55–90. In Computational Intelligence Aided Systems for Healthcare Domain. CRC Press.

Lahat, Dana, Tülay Adali and Christian Jutten. (2015). Multimodal data fusion: an overview of methods, challenges, and prospects. Proceedings of the IEEE 103(9): 1449–1477.

Li, Yi-Ou, Tülay Adali, Wei Wang and Vince D. Calhoun. (2009). Joint blind source separation by multiset canonical correlation analysis. IEEE Transactions on Signal Processing 57(10): 3918–3929.

Lin, Dongdong, Vince D. Calhoun and Yu-Ping Wang. (2014). Correspondence between fMRI and SNP data by group sparse canonical correlation analysis. Medical Image Analysis 18(6): 891–902.

Liu, Jinghui, Daniel Capurro, Anthony Nguyen and Karin Verspoor. (2023). Attention-based multimodal fusion with contrast for robust clinical prediction in the face of missing modalities. Journal of Biomedical Informatics 145: 104466.

Liu, Yu, Xun Chen, Rabab K. Ward and Jane Wang, Z. (2016). Image fusion with convolutional sparse representation. IEEE Signal Processing Letters 23(12): 1882–1886.

Ong, Kok-Leong, Daswin De Silva, Yee Ling Boo, Ee Hui Lim, Frank Bodi, Damminda Alahakoon et al. (2016). Big data applications in engineering and science. Big Data Concepts, Theories, and Applications 315–351.

Putrama, I. Made and Péter Martinek. (2024). Heterogeneous data integration: Challenges and opportunities. Data in Brief 110853.

Qi, Shile, Vince D. Calhoun, Theo GM Van Erp, Juan Bustillo, Eswar Damaraju, Jessica A. Turner, Yuhui Du et al. (2017). Multimodal fusion with reference: searching for joint neuromarkers of working memory deficits in schizophrenia. IEEE Transactions on Medical Imaging 37(1): 93–105.

Rudrapatna, Vivek A. and Atul J. Butte. (2020). Opportunities and challenges in using real-world data for health care. The Journal of Clinical Investigation 130(2): 565–574.

Scheurwegs, Elyne, Kim Luyckx, Léon Luyten, Walter Daelemans and Tim Van den Bulcke. (2016). Data integration of structured and unstructured sources for assigning clinical codes to patient stays. Journal of the American Medical Informatics Association 23(e1): e11–e19.

Shaban-Nejad, Arash, Martin Michalowski and Simone Bianco. (2022). Multimodal artificial intelligence: next wave of innovation in healthcare and medicine. pp. 1–9. In Multimodal AI in Healthcare: A Paradigm Shift in Health Intelligence. Cham: Springer International Publishing.

Srivastava, Nitish and Russ R. Salakhutdinov. (2012). Multimodal learning with deep boltzmann machines. Advances in Neural Information Processing Systems 25.

Sui, Jing, Hao He, Godfrey D. Pearlson, Tülay Adali, Kent A. Kiehl, Qingbao Yu et al. (2013). Three-way (N-way) fusion of brain imaging data based on mCCA+ jICA and its application to discriminating schizophrenia. NeuroImage 66: 119–132.

Tawfik, Nahed, Heba A. Elnemr, Mahmoud Fakhr, Moawad I. Dessouky and Fathi E. Abd El-Samie. (2022). Multimodal medical image fusion using stacked auto-encoder in NSCT domain. Journal of Digital Imaging 35(5): 1308–1325.

Wang, Bo, Aziz M. Mezlini, Feyyaz Demir, Marc Fiume, Zhuowen Tu, Michael Brudno et al. (2014). Similarity network fusion for aggregating data types on a genomic scale. Nature Methods 11(3): 333–337.

Wang, Jiaqi, Hanqi Jiang, Yiheng Liu, Chong Ma, Xu Zhang, Yi Pan, Mengyuan Liu et al. (2024). A comprehensive review of multimodal large language models: Performance and challenges across different tasks. arXiv preprint arXiv:2408.01319.

Waring, Jonathan, Charlotta Lindvall and Renato Umeton. (2020). Automated machine learning: Review of the state-of-the-art and opportunities for healthcare. Artificial Intelligence in Medicine 104: 101822.

Zhang, Yu-Dong, Zhengchao Dong, Shui-Hua Wang, Xiang Yu, Xujing Yao, Qinghua Zhou et al. (2020). Advances in multimodal data fusion in neuroimaging: overview, challenges, and novel orientation. Information Fusion 64: 149–187.

7

Empirical Analysis of AI-enabled Techniques in Modern Healthcare

1. Introduction

Artificial intelligence (AI) has led to the creation of applications, frameworks and pipelines that help to analyze and process data in all domains. In healthcare, AI has brought about a significant shift in how medical data is analyzed, interpreted, and utilized in clinical decision-making. AI applications use various Machine Learning (ML) and deep learning (DL) algorithms to generate accurate predictions from data, automating complex tasks, and mine patterns and insights from the data. These ML and DL algorithms help in better understanding of disease, its diagnosis, and preparation of treatment strategies. AI applications need access to large and complex medical datasets. These datasets will cover data from various sources including the EHRs, clinical test results and medical imaging data. In order to process the data together, huge computational resources are needed to prepare an AI healthcare assistant, that can analyze diverse data formats. The deployment of AI models has made healthcare systems capable of handling the needs of precision medicine. Traditional rule-based systems to AI-driven solutions are now required to transform to a data-driven and personalized approach to healthcare.

There are various AI enabled applications and techniques that can be applied in various cases in healthcare. These intelligent applications are used to enhance efficiency, accuracy, and adaptability in clinical environments. By assessing the advantage AI tools can bring into various cases of healthcare providers, policymakers need to analyze the strengths and limitations of AI models. Effective policy frameworks need to be established to review and standardize the ways in which AI models work and access data for their operations. For example, Convolutional Neural Networks (CNNs) and DL algorithms help to process medical imaging data such as X-rays, Computed Tomography (CT) scans, Magnetic Resonance Imaging (MRI),

DOI: 10.1201/9781032722344-7

Ultrasound scans and other imaging data. These imaging data and application of AI for analysis of the data are common in radiology, dermatology, and ophthalmology (Esteva et al. 2019, Yang et al. 2017). The application of AI has demonstrated success under controlled conditions. Translating the outcomes from the controlled conditions to applications based on real-world clinical environments has practical challenges. An empirical analysis is crucial for identifying and bridging the gaps in applying AI and its impacts in healthcare systems.

With the increase in developments of AI, the potential of AI applications assessing the performance of these models and its efficiency, have to be analyzed for the alignment of these tools with clinical workflows. The development of AI tools has huge data quality requirements, and patient safety standards. In clinical applications the reliability of AI systems should be checked to ensure and maintain quality in various applications. The underlying data needs a significant amount of processing as healthcare datasets are often heterogeneous, sparse, and prone to biases (Suresh and Guttag 2019). The detailed analysis on data quality, data integration capabilities and data preprocessing impact the AI performance of healthcare systems. Constant evaluation of the performance of these systems, provides insights into necessary adjustments for improving the predictions and reliability of the tools, across diverse patient populations.

The success and effective implementation of AI models in clinical applications is dependent on model interpretability and computational efficiency. Another important factor that has a role in determining the success of AI systems is their ability to integrate with existing healthcare infrastructures (Topol 2019). Healthcare professionals also need to trust and understand these systems, especially where patient life is involved. The misinterpretations of the AI outputs can lead to adverse patient outcomes. The performance of AI models needs to be constantly reviewed and improvements must be made in the algorithms, to ensure effective use of these models in the clinical environment to improve transparency, and utility.

In this chapter we systematically review the empirical evidence on AI-enabled techniques and its application in healthcare, covering critical aspects such as data quality, algorithm performance, and clinical applicability. We also discuss different AI tools that are currently being used in healthcare systems. We focus on providing an overview to the ML, DL, NLP, and reinforcement learning applications in healthcare use cases, and the issues with how the quality of healthcare systems impacts the algorithm, and its effectiveness and patient outcomes. The need for robust preprocessing and augmentation techniques are emphasized in the chapter. We also evaluate various metrics used to assess AI algorithms and their performance, providing a comparative analysis of different popular models applied in clinical applications. The clinical applicability of AI tools is also discussed focusing on addressing the operational challenges of integrating AI into healthcare workflows. There are various technical and ethical challenges associated with AI techniques. Here in this chapter, we highlight the limitations, interpretability, privacy concerns, and bias that can occur as part of the healthcare systems (Obermeyer et al. 2019), in detail. The empirical analysis provides an in-depth understanding of various AI-enabled techniques and their capabilities and constraints in modern healthcare systems. We also emphasize the need for the development of effective, transparent, and ethical AI systems that align with healthcare systems that commit to patient safety, clinical efficacy, and equitable access.

2. Analysis of Data Quality in Healthcare

Data quality is an inevitable factor in deciding the effectiveness of AI systems. Systems are trained and work based on various data captured from diverse sources. The systems in healthcare rely on reliable data for generating valid predictions. High-quality data ensures that the AI models can perform better and consistently across diverse patient populations and clinics. The data for healthcare systems are complex and heterogeneous. The data has a combination of structured data from sources like electronic health records (EHRs), and unstructured data from sources like clinical notes, medical imaging, lab results, along with real-time sensor data from wearable devices. The complexity in data modalities presents significant challenges in achieving consistent and reliable high-quality datasets. This impacts the model performance, generalizability, and patient outcomes (Suresh and Guttag 2019).

2.1 Challenges in Healthcare Data Quality

AI systems work based on data provided by the patients and data captured based on various devices and imaging sources. The data quality, heterogeneity and variations in techniques that capture data are some of the main challenges in healthcare systems. Combining approaches to healthcare data also face challenges as the data have to be gathered from various departments, institutions, and geographic locations. This diversity also leads to change in format, structure, and content inconsistencies. For example, imaging data may vary significantly between institutions due to differences in equipment, protocols, and image acquisition settings (Zech et al. 2018). Clinical notes that can give insights on the patient vitals and treatment plans are often manually prepared, and will have subjective variations, inaccuracies, and biases; also, these notes will be case to case and different for each patient. The variation in the data can lead to model biases. AI models trained on data from specific populations/regions will perform better in those populations; there can be performance issues in other contexts. This challenge points out the need to have standardized data formats and guidelines and a common data repository consisting of different sample populations.

Another common issue faced in data of healthcare systems is the requirement for gathering more data. Medical tests and reports are gathered for specific purposes; when creating a comprehensive data center, there is a need for more attributes to be captured. The issues with this are the cost involved for various tests—incomplete tests reports; patients being dropped out from one hospital to others; or omissions in record—create issues in designing a common database for proper model training and development. Missing data can affect the performance of AI models. In supervised ML learning algorithms, proper and clean data is required for training ML models. Data cleaning techniques use methods like imputation—a technique that fills missing values based on statistical methods. The effectiveness of data imputation in the preprocessing stage of the data depends on the data distribution that is used for the training. These data generation methods can introduce biases if not carefully handled (Acuna and Rodriguez 2004).

2.2 Data Preprocessing for Improved Quality

Data preprocessing is a critical step that helps to prepare data from various sources to a common dimension. The preprocessing workflows, enabled to process the data, will help with resolving data quality issues. Data preprocessing steps involve cleaning, normalizing, and transforming raw data into appropriate formats for the data storage and application for analysis. Data cleaning process detects and removes noise, inconsistencies, and eliminates duplicates that can be present inside the data. The noise in the data, such as random errors occurring through sensor readings, will lead to inaccurate predictions. Data processing techniques like filtering and outlier detection helps to reduce noise and increase data reliability.

Data normalization and standardization are preprocessing steps that help to ensure that the data is in comparable dimensions across various data samples. Healthcare data, especially data from medical imaging sources, will get significant variation with intensity, scale, and resolution. Data from imaging machines also differ in formats and dimensions. The unclean and processed data will significantly affect the performance of AI models developed with the data. Normalizing pixel intensity in imaging data and scaling continuous variables ensures that the models are not biased toward specific value ranges. This leads to more consistent data which will be more suitable for analysis. There are various preprocessing steps taken by different approaches for preprocessing the data to fine tune the performance of AI algorithms.

Data augmentation is another approach that is used to generate new data by applying transformations in the existing data, to address data scarcity issues in specific applications to train and build AI models. This process is artificially expanding the datasets through data transformation operations like rotation, translation, or cropping in imaging data. Data augmentation helps to improve the AI models and helps to generalize them for better functioning with unseen data. These methods of data preprocessing are applied in various applications for data preparations and fine tuning them to improve their effectiveness in AI models. This approach of data preprocessing to ensure that the quality of data is maintained is especially helpful when the lack of proper data is absent to build and improve models.

2.3 Importance of Data Quality on Algorithm Performance

AI algorithms performance is dependent on data quality. Data quality is an important factor that impacts the performance of AI algorithms. This data should be sourced from trusted and reliable sources, ensuring its validity and credibility. To improve the performance of AI models high-quality verified data is required, this data can be captured from verified sources with proper care in gathering and fine-tuning data. The high-quality of data ensures accurate representations of patterns from various healthcare data. Poor quality of data introduces biases, decreases model accuracy, and affects the interpretation of the predictions based on the AI models. ML models if they are trained on noisy, imbalanced, and incomplete data have more chance of over-fitting, leading to poor performance when applied on real-world data. For example, Che et al. (2018) highlighted that the missing values in EHR data has significantly affected the performance of predictive models in the Intensive Care Unit (ICU). There is a need for effective preprocessing process flows to fine tune and prepare data for effective analysis.

Data quality is to be ensured for improving the AI model performance and generalization. The data from various populations have to be gathered to prepare a common model to be applied in healthcare applications. The patterns are different with regional and national changes. The changes have to be accommodated and the AI models are to be generalized for long term support and application. Model generalization is the process of creating models that can perform better with respect to different patient populations. The generalized models and algorithms are important in healthcare, where patient demographics, disease progression, and healthcare practices vary widely. High-quality, well-preprocessed data helps robust and adaptable models, minimizing the risk of disparities in care quality due to data inconsistencies.

2.4 Data Quality Standards and Best Practices

In order to improve data quality, there are several steps to be taken care of. Data quality assurance starts from where the data originates, where it is processed and where it is stored. The way in which data is stored plays an important role. There are various standards and frameworks proposed to monitor and streamline data collection, its management, and sharing in healthcare. With the efforts to standardize data, formats like Health Level Seven (HL7) and Digital Imaging and Communications in Medicine (DICOM) for storing medical imaging data are designed. The data standardizing processes ensure data consistency across platforms. Data quality assessments help to verify the data with its completeness, accuracy, and timeliness. Reliable data is an essential component for building healthcare AI projects.

Data quality frameworks that are established also promote the interoperability between standards. This helps in the easy integration of AI applications across different health systems. The interoperability of data ensures AI models can access and process data consistently across platforms. The models built on clean, balanced and verified data enhance the utility and applicability of those models in clinical applications. Implementing best practices for data quality, including data validation, routine quality audits, and adherence to standardized data entry protocols, are critical steps to achieving reliable AI in healthcare.

Figure 1 represents the various data quality issues. The various data quality issues and its impact and solution to address these issues are summarized in Table 1. Ensuring data quality is a crucial factor influencing the effectiveness and safety of AI applications in healthcare. By addressing data heterogeneity, missing values, and noise and implementing best practices for data preprocessing, healthcare organizations can enhance the reliability and accuracy of AI-enabled techniques. High-quality data improves model performance and supports the generalizability and

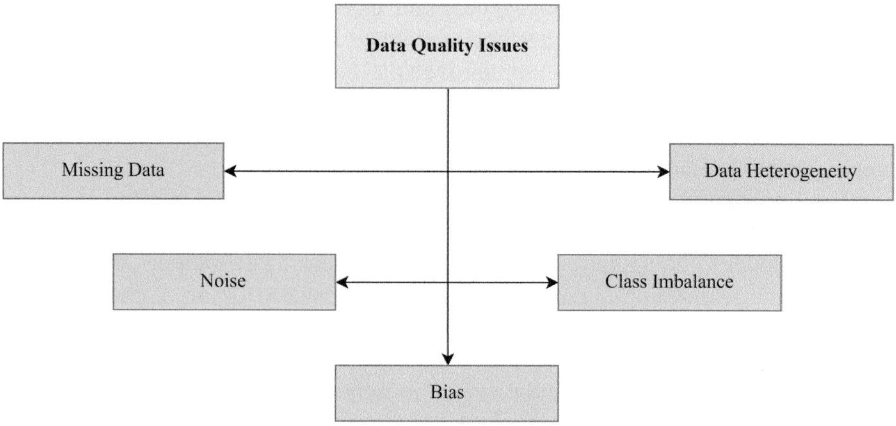

Figure 1. Data quality issues.

Table 1. Data quality issues.

Data Quality Issue	**Description**	**Impact on AI**	**Common Solutions**
Missing Data	Incomplete medical records	Reduces model accuracy	Data imputation
Data Heterogeneity	Variety in data formats	Limits model generalizability	Standardization
Noise	Random errors in data	This leads to inaccurate predictions	Filtering techniques
Class Imbalance	Overrepresentation of certain cases	Affects model performance	Data augmentation
Bias	Non-representative training data	Causes unfair outcomes	Bias mitigation algorithms

fairness of AI systems across patient populations, underscoring the need for rigorous data quality standards in healthcare AI research and development.

3. Analysis of AI-enabled Healthcare Systems

AI-enabled healthcare systems make use of the data and create AI algorithms to make predictions and process medical data. These AI algorithms help to process the data that are interpreted, and with the insights gained, AI algorithms facilitate in making clinical decisions. AI based systems consist of a variety of AI algorithms, used in various capacities for building predictive systems which can model, diagnose and provide personalized treatment assistance. Various AI algorithms and their applications and performance in the use cases are analyzed. The performance evaluation of intelligent techniques is usually done using the evaluation metrics and through computational and interpretative challenges. There are challenges associated with integrating AI models in clinical practice. The evaluation methods and the challenges along with some of the ways to address the challenges are discussed below.

3.1 Evaluation Metrics in AI-enabled Healthcare

Evaluation metrics are used to measure the performance of AI models. There are several measures to monitor the effectiveness and performance of AI models. Commonly used metrics include accuracy, sensitivity, specificity, precision, recall, F1 score, and Area Under the Receiver Operating Characteristic Curve (AUC-ROC). These metrics provide insights on the functioning of the AI algorithms and their performance on the training, testing, validation datasets and functioning on real-time dataset. The accuracy measure represents the proportion of correct predictions made by the model. Sensitivity and specificity measures highlight the model effectiveness in identifying true positive cases and avoiding false positive errors (Chicco and Jurman 2020).

In measuring the performance of healthcare applications, sensitivity is often used. The AI models are used in various applications. One of the main functions of the AI models is disease screening. In disease screening, it is critical to identify the potential cases without leaving the patient at risk. For example, in mammography-based breast cancer screening, high-sensitivity models can help in early detection cases and recommend the treatment outcomes (McKinney et al. 2020). Specificity also helps in diagnostics applications to reduce the unnecessary treatments and interventions. This also helps to minimize the complications in medical decision making.

Evaluation metrics F1 score and AUC-ROC are commonly used to provide a balanced view of a model performance. Depending on the measures, F1 score and AUC-ROC are applied to handle the model performance evaluation, in cases where there is a class imbalance. Lack of data is one of the common challenges in developing AI models. The class imbalance has to be handled for the issues with lack of data and imbalance in the datasets. For example, data where the cases of healthy records may be less than the disease cases. Applying the inputs from multiple metrics helps to validate the observations. AI-enabled healthcare systems are rigorously evaluated, ensuring that models are accurate and clinically applicable for various scenarios.

Table 2 summarizes various AI techniques and their applications in healthcare systems. The AI technique, where it is applied and the benefits and limitations of

Table 2. AI techniques and their applications in healthcare systems.

AI Technique	Application Area	Benefits	Limitations	Example Use Cases
CNN	Medical Imaging	High accuracy in image analysis	Requires large labeled datasets	Detecting tumors in radiology
Random Forest	EHR Analysis	Effective for structured data	Can overfit with small datasets	Predicting patient readmission
NLP	Clinical Documentation	Processes unstructured text data	Challenges in understanding clinical language	Extracting information from notes
Reinforcement Learning (RL)	Personalized Treatment	Learns optimal strategies from the environment	Computationally intensive	Recommending drug dosages

Table 3. Evaluation metrics and clinical applications.

Metric	Definition	Clinical Application	Significance in Healthcare
Accuracy	Proportion of correct predictions	General performance indicator	Basic assessment of model effectiveness
Sensitivity	Ability to detect true positives	Important in disease screening	Ensures high detection of at-risk patients
Specificity	Ability to avoid false positives	Key in diagnostics to prevent false alarms	Reduces unnecessary treatments
F1 Score	The harmonic mean of precision and recall	Balanced metric for imbalanced data	Maintains balanced prediction
AUC-ROC	The area under the ROC curve	Overall model discrimination	Comprehensive evaluation across thresholds

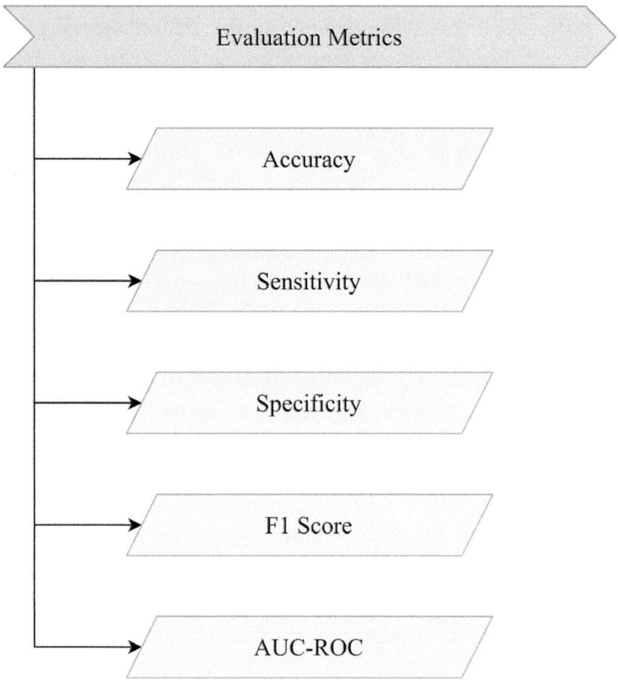

Figure 2. Evaluation metrics.

using the AI tools for the use case are summarized in the table. Table 3 summarizes various evaluation metrics and their applications in building AI models for healthcare systems. Table 3 contains the evaluation metric, definition, clinical application and significance. Figure 2 represents the various evaluation metrics for evaluating the performance of AI models. Understanding the way AI models work and the performance evaluation rubrics helps to identify new applications for the AI tools.

3.2 Performance of AI Algorithms in Different Healthcare Applications

AI algorithms are made to assist healthcare professionals in analyzing data and to make decisions. Various applications of AI systems have significant performance across healthcare applications. The performance of an AI system is influenced by factors like the type of data, target conditions that the model is designed to work in, and the clinical application for which the model is designed. AI systems in healthcare have applications in analyzing medical images and diagnosis processes, predictive analytics using the EHR data, genomics and precision medicine, NLP usage for chatbots and document processing. Table 4 represents the significant impacts through the application of AI systems.

Table 4. Impacts of AI system and its application in healthcare.

Application	**Impacts and Performance of AI algorithms**
Medical Imaging and Diagnostics	Convolutional Neural Networks (CNNs) and other AI models and DL architectures have high accuracy and performance in analyzing medical imaging data. AI models are designed to classify and estimate the regions from medical images. Imaging sources such as radiographs, MRIs, and CT scans are commonly studied. AI algorithms work for diseases including lung cancer, cardiovascular disease, and various neurological disorders. Applications of CNNs have shown diagnostic accuracy to human radiologists in detecting abnormalities in chest X-rays and identifying lesions in MRI scans (Esteva et al. 2019, Yang et al. 2017). There are various other applications of DL models in working with imaging data. Medical imaging and AI models have been designed for stroke, Alzheimer's disease, epilepsy, multiple sclerosis, brain tumors and many other diseases.
Predictive Analytics using EHR Data	Electronic health records (EHR) contain patient data. This data is present in proper format and is documented in healthcare systems. This information can help in building predictive models to predict patient outcomes. The predictions include hospital readmission, sepsis risk, and mortality. ML algorithms, such as logistic regression, random forests, and gradient boosting, are widely used in building predictive models. ML algorithms show promising results in predicting patient deterioration and optimizing treatment plans (Rajkomar et al. 2018). These AI models also use feature engineering to identify the best attributes for the analysis and also use them to extract actionable insights from structured data (e.g., lab results, medication history) and unstructured data (e.g., physician notes), enabling more personalized care.
Genomics and Precision Medicine	AI models, particularly DL networks, are used to analyze genomic data. The application of AI models could easily identify genetic markers associated with diseases. The genetic susceptibility and response to treatment can also be identified using frequent analysis of genomic data. In oncology, AI models help to categorize brain tumors based on genetic profiles. Understanding the genetic profiles leads to personalized therapies to improve patient outcomes (Topol 2019). Integrating federated learning and transfer learning approaches, the systems can build rigorous models that can be easily applied to the context. These AI models can be trained on multi-regional datasets without compromising data privacy, facilitating better generalization in genomics-based predictions (Rieke et al. 2020).
NLP for Clinical Documentation	NLP, which is a subset of AI, has algorithms that help to process unstructured text data. These algorithms enable AI systems to interpret clinical notes, discharge summaries, and medical literature. NLP tools also help to assist with extracting relevant information from vast amounts of clinical documentation, supporting diagnostic decisions, automating report generation, and providing clinical decision support. Techniques like Bidirectional Encoder Representations from Transformers (BERT) and other transformer models have improved the accuracy of NLP tasks, from entity recognition to sentiment analysis, in clinical contexts (Huang et al. 2019).

3.3 Computational and Interpretative Challenges

AI models come with a lot of computational challenges. These models learn from data sources by analyzing the data and extracting features to get hidden patterns from data. AI-enabled systems have shown significant performance with predictive assistance and applications in real-time healthcare systems. There are significant challenges for these systems that include computational efficiency, interpretability, and clinical integration. Table 5 discusses these challenges and how AI systems handle these challenges. Understanding the issues with AI systems helps to minimize the issues with these systems and build more effective systems in future.

Figure 3 demonstrates various challenges with AI models and their integrations with healthcare applications. Table 6 summarizes various challenges in AI models and their integrations in real-time healthcare systems. This gives a clear picture of how the various applications are adopting AI systems and the acceptance of these systems in healthcare.

Table 5. Computational challenges and AI systems.

Challenge	AI Systems
Computational Efficiency	Computational requirements are one of the main challenges of AI algorithms. The AI models need a high amount of data and computational resources for learning from the data and be useful for predictions. The requirement for high computation, storage and data requirements limits their scalability and real-time applicability in healthcare. The computational cost is exceptionally high for models applied to imaging and genomics, where data is voluminous and complex. Techniques like model pruning, quantization, and efficient architectures such as MobileNet are being explored to reduce computational demands, enabling AI systems to be deployed on highly specialized hardware infrastructure (Howard et al. 2017).
Interpretability and Trust	Model interpretability is another challenge that the DL models face. The decisions made by neural networks are not able to be properly interpreted. The features learned for the predictions also are not justifiable in all cases. Especially in healthcare systems where clinicians must understand how a model arrives at a specific conclusion, this model raises challenges. Black-box DL models present challenges with interpretability, the decision-making process in these models are hidden from the normal user and the explainability of the model predictions creates challenges. Explainable AI (XAI) methods, including SHAP (Shapley Additive Explanations) and LIME (Local Interpretable Model-agnostic Explanations), are employed to provide insights into model predictions, allowing clinicians to assess the validity and trustworthiness of AI-driven recommendations (Rudin 2019).
Integration into Clinical Workflows	AI systems often face barriers to integration within clinical workflows due to interoperability issues, differences in healthcare infrastructure, and a lack of standardized protocols. For AI to be truly impactful, it must seamlessly integrate with existing systems, including EHR platforms, imaging tools, and decision support systems. Interoperability standards like FHIR (Fast Healthcare Interoperability Resources) enable AI systems to interact effectively with other healthcare technologies, ensuring that predictions are readily accessible to healthcare providers without disrupting clinical workflows (Bender and Sartipi 2013).

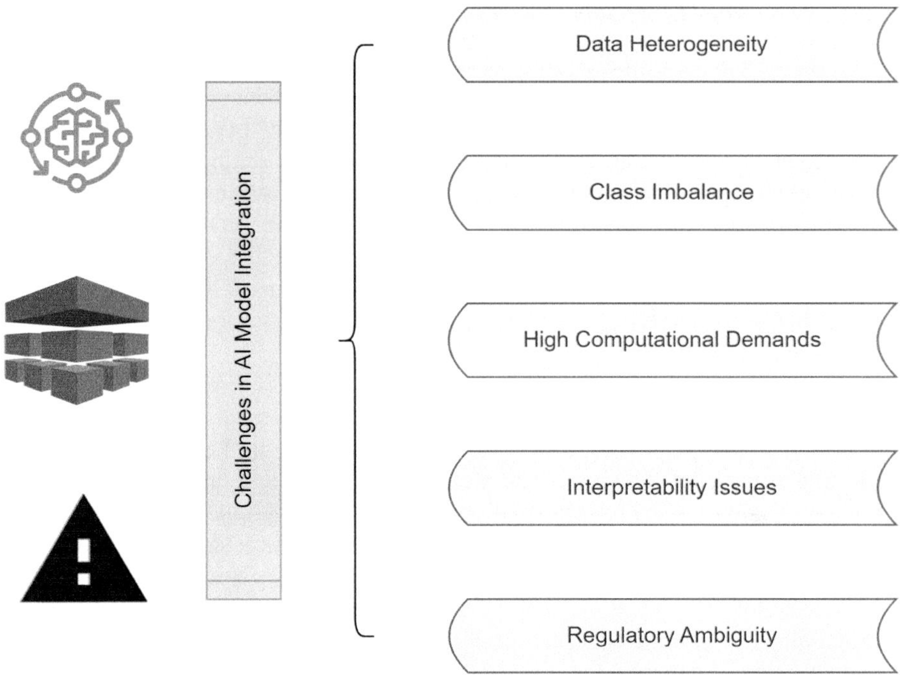

Figure 3. Challenges with AI models and their integration.

Table 6. Challenges in AI model integration.

Challenge	Impact	Proposed Mitigation	References
Data Heterogeneity	Reduces model generalizability	Standardizing formats (DICOM, HL7 FHIR), interoperability frameworks	Bender and Sartipi (2013)
Class Imbalance	Leads to biased predictions	Data augmentation, synthetic data generation	Che et al. (2018)
High Computational Demands	Limits deployment in resource-constrained settings	Model pruning, MobileNet architectures	Howard et al. (2017)
Interpretability Issues	Limits clinician trust	Incorporating explainable AI (SHAP, LIME)	Rudin (2019), Guidotti et al. (2018)
Regulatory Ambiguity	Slows AI adoption	Updated guidelines for Software as a Medical Device (SaMD)	Jiang et al. (2017)

3.4 Case Studies in AI-enabled Healthcare Systems

There are various case studies with AI enabled healthcare systems highlighting the potentials and limitations of AI. DL models for diabetic retinopathy screening have been applied and have shown high sensitivity and specificity in controlled trials. The deployment of models that can be applied on real-world settings has encountered several challenges. The challenges that stop the application of DL models in real-time are due to differences in patient demographics, imaging quality, and clinical workflows (Gulshan et al. 2016). There are several case studies that help to identify the importance and challenges of these systems. The analysis of these systems helps in identifying and addressing the issues with performance, for broader implementation of intelligent systems in healthcare.

AI models developed for sepsis prediction in intensive care units have high accuracy in early detection; however, studies have revealed that these models can produce false positives, leading to alarm fatigue among healthcare providers. Adjusting the model thresholds and designing ensemble techniques are ongoing strategies to improve the reliability of these AI models (Desautels et al. 2016). The deployment of AI models should not affect the roles of clinical staff, in fact, these AI tools should be an assistant for them to get a border picture about the patient.

AI-enabled healthcare systems have huge potential and they can be applied in domains like diagnostics and predictive analytics to genomics and clinical documentation. Realizing the full potential of AI systems and addressing the computational and interpretive challenges is crucial with healthcare applications. Aligning AI models and their outputs with clinical requirements and expectations is another important factor that is to be taken care of. Through empirical evaluations of the system performance, with evaluation metrics and real-world case studies, the healthcare sector can take advantage of the AI systems and accurate, efficient, and seamlessly integrated AI tools can be designed for clinical environments.

4. Clinical Applications and AI-Enabled Techniques

AI-enabled techniques help in assisting clinicians in offering better and personalized patient care. AI tools help in the design and development of advanced tools for diagnosis, risk stratification, treatment planning, and patient monitoring. For the design and development of tools, AI algorithms from ML, DL, and NLP are applied in the design and development of applications. These applications help to process the data from imaging, genomics, real-time decision support and personalized medicine. This section discusses various AI applications that are used in clinical settings, focusing on specific applications, the effectiveness of different techniques, and emerging challenges.

4.1 Diagnostic Imaging and Radiology

AI models are designed to process medical imaging—the models are able to process and detect the disease from the medical images. The application of medical image processing is one of the most significant advancements in healthcare. For medical image processing DL algorithms, especially Convolutional Neural Networks (CNNs), are used. AI models can also detect anomalies in X-rays, CT scans, MRIs, and other imaging modalities. The AI based image processing helps to achieve more diagnostic accuracy comparable to or exceeding that of human radiologists in several applications. CNNs are effective in detecting lung nodules, identifying diabetic retinopathy in retinal images, and classifying skin lesions (Esteva et al. 2019, Yang et al. 2017).

In radiology, AI models help to reduce diagnostic errors made by human agents. AI models also help with the early detection of diseases and improve patient outcomes. The Food and Drug Administration (FDA) has approved the use of AI models in processing medical imaging. AI diagnosis of imaging modalities helps in recognizing the abnormalities in the imaging data and assists radiologists in detecting breast and lung cancer. The main challenge for the application of AI models is to make the models work in a generalized manner, for application in diverse patient populations. The imaging quality, variation in equipment and protocol impacts the model performance (Zech et al. 2018). The black box nature of DL models also raises concerns about interpretability and trust in medical decision-making.

4.2 Predictive Analytics and Risk Stratification

Predictive analytics systems help in assessing patient risks. With the engine being trained on various features, the systems are effective in analyzing and processing patient data effectively. These AI enabled systems enable proactive interventions for the patients. The predictive models are applied for analyzing patient data predicted for hospital readmissions, assessment of sepsis risk, and identification of patients who are at risk of deterioration. ML algorithms like logistic regression, decision trees, and ensemble methods like random forests, are used in predictive modeling with EHR data. The designed intelligent systems help to provide personalized care for the patients by assessing and analyzing their real-time conditions.

AI models deployed in intensive care units (ICUs) help to predict sepsis. This helps in early detection and timely intervention. The early detection and interventions help to reduce mortality rates (Desautels et al. 2016). Models analyze and monitor patients in real-time and give alerts for expert medical attention. The AI models analyze patient data in real-time to obtain insights from various data sources. AI analyze and process data from various sources including lab results, vital signs, and clinical notes. AI models predict sepsis risk with high sensitivity. Predictive models are used to identify cardiovascular risk, assess mortality risk, and determine patient discharge planning, enhancing safety and resource allocation in healthcare systems (Rajkomar et al. 2018).

4.3 Genomics and Personalized Medicine

AI is applied in applications to process genomics data enabling personalized medicine. The targeted approach to treatment by analyzing the individual's genetic profiles help to address the cause for the disease. ML models help to identify the genetic markers that are responsible for various conditions and predict drug responses. AI models can classify tumors based on molecular characteristics and help to detect the actual factor that is responsible for the tumors. In oncology DL models help to analyze genomic data to identify cancer subtypes, helping clinicians determine the most effective treatment for individual patients (Topol 2019).

Transfer learning is an AI approach that is used to train AI models on large and diverse datasets. Training on diverse and large datasets allows the models to improve their learning and predictive capabilities and to generalize across populations. The models learn from different data and the models are updated accordingly to perform well with diverse data sources. The transfer learning approach for training AI models is useful when designing models for rare diseases and complex conditions where data scarcity is a problem. In federated learning, the models are trained on decentralized data without compromising patient privacy. This helps in addressing one of the major concerns in genomics (Rieke et al. 2020). Along with the advancements with AI techniques, the challenges of the AI models remain with validating the AI models and their predictions. The model's predictions need to be evaluated across diverse patient populations and ensure that the predictions made are helping the patients to improve clinical outcomes.

4.4 Natural Language Processing (NLP) for Clinical Documentation and Decision Support

NLP, a subset of AI, is an essential tool in clinical applications to gather data and interpret them from unstructured data sources. NLP and its derivatives can process and analyze data from clinical notes, discharge summaries, and medical literature. AI models power NLP algorithms to extract information from text documents, for expert assistance with diagnostic decisions, automating report generation, and supporting clinical decisions. NLP algorithms process EHR data enabling these models to assess patient medical history and symptoms and assist clinicians with real-time recommendations (Huang et al. 2019).

The NLP based chatbots and virtual assistants are integrated into healthcare systems by providing access to patient information. These AI models are used to support the patients and clinicians. These intelligent tools answer common patient queries, assist in booking appointments, offer mental health support, reduce workload and assist the healthcare staff. Transformer models including Bidirectional Encoder Representations from Transformers (BERT) have greatly improved the accuracy of NLP applications, enhancing tasks such as entity recognition, summarization, and sentiment analysis in clinical applications (Kenton et al. 2019). AI and NLP have huge roles with the design and development of AI applications to help the patients in hospital and assist and monitor them for better patient care.

4.5 Real-Time Monitoring and Patient Management

AI models are applied for real-time patient monitoring. For real-time patient monitoring, the AI models access data from wearable devices and IoT-enabled healthcare tools. These models continuously analyze physiological data, such as heart rate, oxygen levels, and activity patterns, to monitor chronic conditions, detect anomalies, and provide alerts to clinicians. AI algorithms connected to wearable devices are used to monitor cardiac conditions. These tools can notify healthcare providers about potential arrhythmias in real-time. The tool is connected to inform the people associated with immediate intervention for the user of the wearable device.

AI-enabled glucose monitoring systems analyze blood glucose levels to assist patients with diabetes. These intelligent tools assist the doctors with insulin dosages, and provide dietary recommendations. AI-driven mobile apps track behavioral patterns, providing insights into patient mood and potential triggers in monitoring mental health of participants. These intelligent apps help in early detection and assistance for mental health conditions including depression, anxiety and so on. AI enabled tools allow patients to manage their health and support clinicians monitoring patients remotely. The application of intelligent tools offering personalized management plans for patients.

4.6 Challenges in Clinical Applications of AI

AI has many applications in healthcare; deploying applications with AI for clinical use cases have challenges. The AI tools use trained DL models on various datasets; the interoperability of these models is a critical issue. When applying AI models for different use cases, the clinicians should be able understand how the AI tool could give the suggestion. Explainable AI (XAI) methods have been used to explain the predictions by the AI models. XAI techniques like Shapley Additive Explanations (SHAP) and Local Interpretable Model-Agnostic Explanations (LIME), are used to improve the explainability issues of AI models. Improvements on the current techniques and improving the explainability of the AI models is required to make these tools more reliable and trustworthy for clinicians (Rudin 2019).

Clinical applications of AI should also address the regulatory requirements to ensure patient safety and data privacy. Ensuring that the applications are compliant with frameworks like the Health Insurance Portability and Accountability Act (HIPAA) and General Data Protection Regulation (GDPR). Compliance with standards and frameworks are necessary to secure data handling and patient trust. Technical limitations such as computational demands and interoperability challenges also affect the phase of AI integration in clinical workflows. There is a need for streamlined and scalable models (Bender and Sartipi 2013), that are responsible and explainable.

AI-enabled techniques have significant advancements in clinical applications. These intelligent tools are used for diagnostics, predictive analytics, personalized treatment and real-time monitoring. These applications are used to improve diagnostic accuracy, helping with personalized treatment planning, and support continuous patient monitoring. The AI tools are aimed to provide better patient outcomes and efficient resource allocation in medical facilities. Achieving widespread clinical adoption of AI tools requires solving the challenges with model interpretability, data privacy, and system integrations. With the advancements in AI technology and regulatory compliance, AI-enabled healthcare applications are used to transform clinical practice, setting new patient care and management standards.

5. Challenges and Limitations

AI-enabled medical tools have huge potential in transforming and assisting healthcare professionals in providing better patient care. There are challenges associated with the use of these tools. These limitations reduce the acceptance of the AI applications and there is a need to ensure effectiveness, reliability and ethical use of these tools in clinical practice. These challenges of these tools include technical limitations, ethical concerns, regulatory compliance, and issues related to data quality and accessibility. Here we discuss and summarize key challenges and constraints that influence AI performance, integration, and societal impact in healthcare.

5.1 Technical Challenges

Some of the major technical challenges with respect to the application of AI enabled techniques in modern healthcare are discussed in Table 7. The key challenges include the issues with model interoperability and explainability, generalization and transferability and the computational requirements and issues with the scalability of these models.

Table 7. Technical challenges.

Model Interpretability and Explainability	Model interpretability is a critical challenge in healthcare AI systems. The black-box nature of the DL algorithms affects the interpretability. There is a need that the clinicians and healthcare providers must understand the reason for a recommendation from an AI system. Complex AI architectures including the neural networks, needs more transparency, making it easier to interpret their decision-making process (Rudin 2019). XAI techniques such as SHAP and LIME, aim to enhance interpretability by highlighting the features contributing to specific predictions. More research is required to develop user-friendly interpretability tools that facilitate AI integration into clinical workflows.
Generalizability and Transferability	AI models are trained on specific datasets. They are expected to predict and work with similar data sources. These models won't be able to perform as expected for diverse patient populations, healthcare settings, and geographic regions. This is due to the issues and differences with data sources. Factors like demographic variations, differences in clinical practices, and data acquisition protocols also lead to the development of biased and non-generalizable models. To ensure that AI systems are adaptable to various clinical environments training of AI models on diverse, representative datasets is required. Along with large curated data sources, techniques like transfer learning help the AI models to adapt to new contexts (Zech et al. 2018).
Computational Requirements and Scalability	AI tools used for processing the medical imaging and genomics data requires a huge number of computational resources for training, development and deployment. The requirement of high computational requirements limits their application in resource-constrained use cases. The hospitals or clinics with limited infrastructure will not be able to make use of the advanced AI tools as they require high computational resources. Techniques like model pruning, quantization, and use of lightweight architectures (e.g., MobileNet) are being explored to reduce computational demands while achieving real-time performance on limited hardware remains challenging (Howard et al. 2017).

5.2 Data Quality and Availability

Another main challenge with the AI techniques application is data quality and availability. Table 8 summarizes some of the major challenges with respect to gathering quality data and its availability. Key factors associated with data quality and availability include data scarcity and imbalance in the available data. Data privacy and security is another challenge. The data heterogeneity and standardization of the available data is another challenge with the use of AI tools.

Table 8. Data quality and availability issues.

Data Scarcity and Imbalance	High quality data is a major requirement with training AI models. The data availability for healthcare systems are limited. When it comes to rare diseases, the models will not be able to identify the disease and its symptoms, due to lack of learning from the data for the disease. In medical datasets, the data will be imbalanced in most of the cases. Strategies have to be devised to ensure that the over representation of features or the under representation of features does not affect the performance of AI systems. Manual intervention is required in these cases. Data augmentation techniques, such as synthetic data generation, address these issues, yet they have limitations and can sometimes introduce biases if not implemented carefully (Che et al. 2018).
Data Privacy and Security	AI systems require access to patient information. These details include personal data, which raises privacy and security concerns. Data access should be based on compliance protocols and regulations. Health Insurance Portability and Accountability Act (HIPAA) and General Data Protection Regulation (GDPR) are regulations that should be considered while designing AI systems. It is required to ensure that the patient's trust is kept and unauthorized data access to their data is prevented. Federated learning and differential privacy are techniques that allow models to learn from decentralized data without compromising privacy (Rieke et al. 2020).
Data Heterogeneity and Standardization	Data comes from various sources; it includes EHRs, medical imaging, genomics, wearable devices, documents, test results and so on. Each of these data differ in format, quality, and structure. This heterogeneity of the data complicates the data integration process. There is a need for developing comprehensive and generalizable models for data processing and aggregation. Standardized data formats (e.g., FHIR for EHRs, DICOM for imaging) and ensuring interoperability across systems is critical for integrating AI techniques in healthcare (Bender and Sartipi 2013).

5.3 Ethical and Social Challenges

Ethical and social challenges are also associated with the development and application of AI techniques in modern healthcare systems. Table 9 summarizes few of the ethical and social challenges that exist with AI systems. Some of the major issues include the biases and lack of fairness with respect to algorithmic systems; DL models lack transparency and accountability. These intelligent systems need patient autonomy and informed consent when processing their data.

Table 9. Ethical and social challenges.

Algorithmic Bias and Fairness	AI models can carry biases from the data that they have been trained in. Biases in the models will cause incorrect predictions and conclusions. Models trained on data from specific populations may not perform well with people from demographics. Ethical guidelines and bias detection techniques are necessary to ensure that AI models are fair across all patient populations (Obermeyer et al. 2019). Ensuring that biases are not affecting the model predictions is one of the crucial steps to make AI responsible.
Transparency and Accountability	Determining who will be taking the accountability for the decisions along with AI-driven decisions in healthcare has ethical and legal challenges. When an AI model provides a recommendation that leads to an adverse outcome, who will be taking the responsibility is the question. Establishing transparency in AI processes, including documenting model development and training methodologies, is essential to solve these issues. Regulatory bodies increasingly emphasize accountability frameworks, but implementing them effectively requires clear policies and collaboration across disciplines (Topol 2019).
Patient Autonomy and Informed Consent	AI applications in healthcare have concerns about patient autonomy and consent process. P must be aware about AI and its role in clinical decision making. Before starting the treatment, patients need to fully understand how AI models can contribute to their diagnoses and treatment plans. This helps to ensure that the patients are informed about the role of AI and the advantages and challenges in their care. If they are not interested, they can opt out of AI application and assistance. Healthcare providers should balance AI recommendations with patient preferences, and help to make a collaborative approach to AI-driven care (Gulshan et al. 2016).

5.4 Regulatory and Legal Challenges

Regulatory and legal challenges are associated with the implementation of AI systems. Policy makers need to create strong policy frameworks that help to evaluate and monitor the application of AI systems. Some of the challenges for AI systems include the challenge to meet the standards set as part of the compliance and regulatory frameworks, validation and approval process, intellectual property and data ownership. These challenges and concerns have to be addressed in order to make the AI systems efficient and reliable. Some of the major regulatory and legal challenges are mentioned in Table 10.

AI-enabled healthcare techniques have various challenges and it is required to address them before the real-time application of these tools. The challenges can impact their implementation, performance, and ethical implications. Addressing the limitations of AI models requires collaboration across technical, clinical, and regulatory disciplines to develop solutions that balance innovation with patient safety and ethical responsibility. Technical advancements are required for improving the interpretability of the model, data privacy, regulatory compliance, and ethical guidelines. Fairness and transparency are essential to realize the full potential of AI in healthcare. As AI continues to evolve, an interdisciplinary approach is required to overcome the challenges and facilitate responsible and equitable integration of AI and its applications into clinical practice.

Table 10. Regulator and legal challenges.

Compliance with Regulatory Standards	Compliance with regulatory standards is a requirement for systems and applications. AI enabled systems are used for predictive analytics work based on data. The AI models learn and work, based on the data and the learning they made from the data, to provide recommendations. The AI systems must follow the stawndards to ensure that patient safety and data privacy is taken care of. Regulations such as the HIPAA and GDPR need to be adapted to the complexities of AI technologies. The compliance should remove the ambiguity around issues like data sharing, model validation, and patient consent. Regulatory bodies are gradually developing guidelines for AI in healthcare. Consistent and comprehensive policies are required to handle ethical and technical concerns (Jiang et al. 2017).
Validation and Approval Processes	Development of AI models should be followed by rigorous testing and validation before being approved and applied for clinical use. This lengthy, resource intensive and challenging task is essential to ensure the systems performance. AI algorithms require continuous updates and adjustments, making regulatory approval, an ongoing process. The Food and Drug Administration (FDA) and other regulatory bodies have started to establish frameworks for AI model validation. The Software as a Medical Device (SaMD) guidelines is an example for this framework; adapting to these frameworks to address the unique, evolving nature of AI in healthcare is still a challenge.
Intellectual Property and Data Ownership	The ownership of AI-generated insights, recommendations based on the patient data is an issue in healthcare systems. Doctors who are treating the patients should use these recommendations as a suggestion and this should not divert treatment plans and the blame put on the system. Issues also arise around the proprietary nature of some AI models, where model developers may restrict access to the model's functioning or underlying algorithms. Establishing clear legal frameworks for data and model ownership is essential to ensure fairness and transparency in AI-driven healthcare.

6. Ethical Considerations

Integrating AI in healthcare systems has various ethical issues. It is important to resolve these issues to ensure that the tools are acting responsible. AI systems can transform the healthcare systems by improving the diagnostic accuracy, personalized treatment, and enhanced patient care. The system also has risks related to privacy, fairness, transparency, and accountability. Ethical principles in AI-driven healthcare systems are critical in giving the vulnerable nature of patient populations and the high stakes of clinical decisions. Some of the major ethical issues have been summarized in Table 11, where the issue and the existing solutions have been detailed Dwork and Roth (2014), Rieke et al. (2020), Obermeyer et al. (2019), Topol (2019), Rudin (2019), Guidotti et al. (2018), Jiang et al. (2017), Gulshan et al. (2016), Faden and Beauchamp (1986).

Table 11. Ethical considerations in AI for healthcare.

Ethical Issue	Description	Proposed Solution
Privacy and Data Security	Protecting sensitive patient information from breaches or misuse	Federated learning, differential privacy
Algorithmic Bias	Unequal model performance across demographic groups	Bias detection, diverse training datasets, fairness-aware algorithms
Transparency and Explainability	Difficulty in understanding complex AI decision-making	Explainable AI methods like SHAP, LIME
Accountability	Determining liability for adverse outcomes	Clear documentation, shared responsibility between developers and clinicians
Patient Autonomy	Ensuring patients understand AI's role in their care	Transparent consent processes, patient education

6.1 Privacy and Data Security

One of the major ethical concerns is ensuring the privacy and security of sensitive patient data. Misuse of patient data can be done by many people who are involved in providing insurance, preparing medicines and so on. Hence, data security and privacy are major factors that have to be taken care of. Healthcare data contains personal and sensitive information, which includes diagnoses results, treatment histories, and genetic profiles. The HIPAA and GDPR frameworks help to enforce data protection standards. To monitor and ensure that the systems comply with the frameworks fully and with the regulations is also challenging. This happens when the data needs to be shared across institutions to build complex and scalable systems (Rieke et al. 2020).

Federated learning and differential privacy improve data security. Federated learning is a learning mechanism in which AI models are trained on decentralized data sources; and differential privacy introduces controlled noise into datasets, making it possible to extract patterns without revealing individual patient identities. Implementing these techniques will safeguard patient privacy while enabling data-driven insights (Dwork and Roth 2014).

6.2 Algorithmic Bias and Fairness

Algorithmic bias is an important ethical issue that has to be taken care of. AI models, if not monitored properly, can work based on the disparities in them; this might be captured from selection of wrong data or the population. Bias can also be introduced to the AI models from non-representative training data or issues with model design. This leads to unequal model performance for different demographics. AI models trained on data from specific ethnic or socioeconomic groups perform on populations not well-represented in the data (Obermeyer et al. 2019). If the data used for training the ML model misses to represent a portion from the population; there will be issues with predictions and recommendations made for those missed populations.

In order to address the bias issues with AI systems, it is necessary to ensure that datasets used for training healthcare AI systems are diverse and represent all populations. Specific care needs to be taken to address the bias, using bias detection and mitigation algorithms. These algorithms help to identify and reduce model bias and ensure fairness during training. Ethical frameworks and guidelines help to prioritize equity, reduce disparities and bias in predictions. Removal of bias from the systems helps the patient to be treated equally and ensures that the AI recommendations do not carry bias.

6.3 Transparency and Explainability

The black-box nature of AI models, specifically the DL models, raises concerns about their transparency and interpretability. When applied for real-time healthcare applications where clinical decisions often have life-altering impacts, clinicians should be able to understand the reasoning behind AI recommendations. Lack of interpretability affects the clinical trust and complicates the process, and patients feel uneasy about decisions driven by opaque algorithms (Rudin 2019).

XAI is a rapidly growing company that aims to make AI models more interpretable by providing insights into how predictions are made. Methods like Shapley values, Local Interpretable Model-Agnostic Explanations (LIME), and Layer-wise Relevance Propagation (LRP) give transparency by identifying key features influencing model outputs. XAI will also face challenges with identifying interpretability in complex DL models. For AI to be ethically implemented in healthcare, more significant efforts are needed to make AI systems transparent and to develop interpretability methods that clinicians can rely on when making patient-care decisions (Guidotti et al. 2018).

6.4 Accountability and Liability

The integration of AI in healthcare introduces complex questions about its accountability and liability. When AI-driven recommendations result in adverse outcomes, it is often unclear who should be held responsible. When applying AI models in healthcare applications, it is good that explainability also comes so that the clinicians can understand the reason for AI recommendations to make wise calls. Determining liability is further complicated because AI models continue to evolve and learn from new data, which can change the model's behavior over time. Ensuring live updates on models is also required to improve their performance based on data and feedback.

Ensuring accountability for AI models requires setting up guidelines and documentation. These guidelines help to track the model's development, deployment, and application. Regulatory frameworks set the importance of accountability in healthcare AI. Initiatives should be taken to promote transparency and adherence to ethical standards. Developers and healthcare providers should ensure that AI-driven decisions are supervised. Clinicians should retain ultimate authority in decision-making to maintain accountability when the use of AI tools (Topol 2019). Having a monitoring system that helps to track the model's performance is essential to address liability concerns.

6.5 Patient Autonomy and Informed Consent

AI applications in healthcare have raised questions on patient autonomy; there are high chances that the patients feel that their treatment decisions are influenced by algorithms rather than human judgment. Respect for patient autonomy ensures that patients understand the role of AI in their care and are informed of how their data is being used. Informed consent processes will give a transparent communication on how AI will impact diagnostic and treatment decisions. This allows patients to make choices regarding their care (Faden and Beauchamp 1986).

Including patients, while taking decisions through the application of AI, makes the process a shared decision-making one. The shared responsibility is a vital aspect of ethical AI integration. Clinicians should discuss the recommendations from AI systems with patients, explaining the potential benefits and challenges. A collaborative approach helps to keep patient trust. This also ensures that AI is used as a supportive tool for patient care. Respecting patient preferences and actively involving them in their treatment decisions aligns with the core ethical principle of patient autonomy and reinforces trust in AI-driven healthcare.

6.6 Ethical Use of Predictive and Preventive AI Applications

AI models are applied to get predictive and preventive recommendations from AI applications. The predictive models are used to identify patients at risk for certain diseases, helping to provide early intervention and preventive measures. Ethical challenges occur when using predictive analytics to make preventive recommendations. A patient, when informed that they have a high likelihood of developing a chronic condition, may experience anxiety that affects their behavior in potentially harmful ways based on algorithmic predictions.

The ethical use of predictive AI requires a balance between actionable insights and avoiding unintended consequences. Healthcare providers should communicate the predictive information to patients by respecting patient well-being. This helps to ensure that the recommendations are actionable, supportive, and free from pressure. Ethical guidelines emphasize that predictions are probabilities rather than certainties. There is a need for reinforcement through careful interpretation and personalized counseling.

6.7 Regulatory Oversight and Ethical Governance

Due to the rigorous pace of AI developments, the technology and its applications have a wide role to play in the daily life of people. In healthcare systems AI is used to automate, analyze and monitor patients. This emphasizes the need for robust regulatory and ethical governance frameworks. Regulatory bodies work together to ensure that the AI systems are following guidelines and mechanisms to address the challenges posed by AI systems. Ethical oversight committees and institutional review boards (IRBs) play an essential role in reviewing and approving AI applications. Review and approval processes make sure that AI applications prioritize moral principles of beneficence, non-maleficence, justice, and respect for autonomy (Jiang et al. 2017).

Ethical governance in AI healthcare projects, involves continuous monitoring and auditing. AI models need periodic assessments to ensure they remain fair, transparent, and effective over time. Incorporating AI ethics experts into project teams and conducting regular audits for bias and performance, will ensure that the systems are efficient. Involving diverse stakeholders for ethical discussions is also a practice that can support responsible and accountable AI deployment in healthcare.

For the safe integration of AI into healthcare there is a need to prioritize and address ethical considerations and challenges. Handling ethical challenges includes forming the frameworks to manage patient privacy, fairness, and transparency in AI systems. The users should be aware about the suggestions and the way their data is processed. AI systems should not block patient rights or misinterpret medical data and provide observations. Frameworks and checkpoints are to be formed to ensure that the AI systems are working as expected. It is also important that AI systems give options for the patient to make their own choices about their care. Doctors, AI developers, and regulators must work together to ensure the responsible deployment of AI systems.

7. Conclusion

The integration of AI solutions in healthcare systems transforms patient care. AI systems are used to assist in clinical practice to analyze data, predict patient outcomes, recommend personalization treatment, and improve operational efficiency. There are various applications for the application of AI tools. The technical challenges and ethical implications of these systems have to be solved to ensure their usability and acceptance. To ensure AI tools are safe to relay, careful consideration and planning are required to enable constant monitoring and evaluation. AI has significant power in diagnostic approaches, processing medical imaging data, creating predictive analytics based on real-time data, genomics data, and patient monitoring. In order for the widespread adoption of AI tools, planning, evaluation and regulatory compliance is required.

The ability of AI systems to enhance diagnostic accuracy, reduce and forecast risk, and manage chronic conditions through predictive models signifies its value as a transformative tool in healthcare. AI models built on DL algorithms, like CNNs for processing medical imaging, and ML to process EHRs, have gained accuracy levels similar to the predictions and observations by clinical experts in certain use cases. AI models, if not trained to work on a generalized and standardized data, often have practical application challenges due to data quality, computational demands, and variability of data standards. The challenges for the AI applications point to the need for robust, standardized data and efficient AI models that are applicable for healthcare applications.

Some of the key challenges like data privacy, model interpretability, and algorithmic bias need to be addressed for the ethical integration of AI. To ensure data security, handling biases in model training and creating interpretable AI systems is necessary to improve clinical trust and protect patient rights. AI systems are applied for clinical decision-making. Accountability and transparency in AI-driven recommendations are essential for maintaining professional and public confidence.

Involvement of healthcare providers helps in effective planning, designing and deployment of AI models. The involvement of experts helps with better understanding of patient needs, clinical workflows, and the ethical complexities of medical practice.

The ethical considerations that should be addressed include privacy, fairness, transparency, accountability, and patient autonomy. Addressing ethical concerns is necessary for the responsible development and deployment of AI systems in healthcare. AI systems used for healthcare applications should aim at the improvement of patient care with precise treatment approaches, by giving importance to patient values and rights. Emerging solutions help to solve the problems with building AI models with federated learning for privacy-preserving data sharing, explainable AI for model transparency, and bias mitigation techniques, aimed at ensuring equitable care delivery. Establishing clear regulatory guidelines and ethical governance frameworks will also be essential for maintaining ethical standards as AI technology evolves.

The future of AI in healthcare lies with a balanced approach that combines technological innovation with ethical responsibility. Interdisciplinary collaboration is required to understand and make use of the potential of AI to solve challenges and innovate solutions in healthcare applications. As AI keeps evolving, new innovations help to ensure human life can be made easy. AI solutions have been applied to improve patient-centered care, enforce ethical standards, and rigorous validation to provide better treatment. Constant monitoring and checks with regulatory frameworks ensures that the AI-enabled systems contribute to society, opening the way for more efficient, and accessible healthcare solutions.

References

Acuna, Edgar and Caroline Rodriguez. (2004). The treatment of missing values and its effect on classifier accuracy. pp. 639–647. In Classification, Clustering, and Data Mining Applications: Proceedings of the Meeting of the International Federation of Classification Societies (IFCS), Illinois Institute of Technology, Chicago, 15–18 July 2004. Berlin, Heidelberg: Springer Berlin Heidelberg.

Bender, Duane and Kamran Sartipi. (2013). HL7 FHIR: An Agile and RESTful approach to healthcare information exchange. pp. 326–331. In Proceedings of the 26th IEEE International Symposium on Computer-based Medical Systems. IEEE.

Che, Zhengping, Sanjay Purushotham, Kyunghyun Cho, David Sontag and Yan Liu. (2018). Recurrent neural networks for multivariate time series with missing values. Scientific Reports 8(1): 6085.

Chicco, Davide and Giuseppe Jurman. (2020). The advantages of the Matthews correlation coefficient (MCC) over F1 score and accuracy in binary classification evaluation. BMC Genomics 21: 1–13.

Desautels, Thomas, Jacob Calvert, Jana Hoffman, Melissa Jay, Yaniv Kerem, Lisa Shieh, David Shimabukuro et al. (2016). Prediction of sepsis in the intensive care unit with minimal electronic health record data: a machine learning approach. JMIR Medical Informatics 4(3): e5909.

Dwork, Cynthia and Aaron Roth. (2014). The algorithmic foundations of differential privacy. Foundations and Trends® in Theoretical Computer Science 9(3–4): 211–407.

Esteva, Andre, Alexandre Robicquet, Bharath Ramsundar, Volodymyr Kuleshov, Mark DePristo, Katherine Chou et al. (2019). A guide to deep learning in healthcare. Nature Medicine 25(1): 24–29.

Faden, Ruth R. and Tom L. Beauchamp. (1986). A History and Theory of Informed Consent. Oxford University Press.

Food and Drug Administration. (2019). Proposed regulatory framework for modifications to artificial intelligence/machine learning (AI/ML)-based software as a medical device (SaMD).

Guidotti, Riccardo, Anna Monreale, Salvatore Ruggieri, Franco Turini, Fosca Giannotti and Dino Pedreschi. (2018). A survey of methods for explaining black box models. ACM Computing Surveys (CSUR) 51(5): 1–42.

Gulshan, Varun, Lily Peng, Marc Coram, Martin C. Stumpe, Derek Wu, Arunachalam Narayanaswamy, Subhashini Venugopalan et al. (2016). Development and validation of a deep learning algorithm for detection of diabetic retinopathy in retinal fundus photographs. Jama 316(22): 2402–2410.

Howard, Andrew G. (2017). Mobilenets: Efficient convolutional neural networks for mobile vision applications. arXiv preprint arXiv:1704.04861.

Huang, Kexin, Jaan Altosaar and Rajesh Ranganath. (2019). Clinicalbert: Modeling clinical notes and predicting hospital readmission. arXiv preprint arXiv:1904.05342.

Jiang, Fei, Yong Jiang, Hui Zhi, Yi Dong, Hao Li, Sufeng Ma et al. (2017). Artificial intelligence in healthcare: past, present and future. Stroke and Vascular Neurology 2(4).

Kenton, Jacob Devlin Ming-Wei Chang and Lee Kristina Toutanova. (2019). Bert: Pre-training of deep bidirectional transformers for language understanding. In Proceedings of naacL-HLT 1: 2.

McKinney, Scott Mayer, Marcin Sieniek, Varun Godbole, Jonathan Godwin, Natasha Antropova, Hutan Ashrafian et al. (2020). International evaluation of an AI system for breast cancer screening. Nature 577(7788): 89–94.

Obermeyer, Ziad, Brian Powers, Christine Vogeli and Sendhil Mullainathan. (2019). Dissecting racial bias in an algorithm used to manage the health of populations. Science 366(6464): 447–453.

Rajkomar, Alvin, Eyal Oren, Kai Chen, Andrew M. Dai, Nissan Hajaj, Michaela Hardt, Peter J. Liu et al. (2018). Scalable and accurate deep learning with electronic health records. NPJ Digital Medicine 1(1): 1–10.

Rieke, Nicola, Jonny Hancox, Wenqi Li, Fausto Milletari, Holger R. Roth, Shadi Albarqouni et al. (2020). The future of digital health with federated learning. NPJ Digital Medicine 3(1): 1–7.

Rudin, Cynthia. (2019). Stop explaining black box machine learning models for high stakes decisions and use interpretable models instead. Nature Machine Intelligence 1(5): 206–215.

Suresh, Harini and John V. Guttag. (2019). A framework for understanding unintended consequences of machine learning. arXiv preprint arXiv:1901.10002 2(8): 73.

Topol, Eric J. (2019). High-performance medicine: the convergence of human and artificial intelligence. Nature Medicine 25(1): 44–56.

Yang, H. Mehta, Duan, T., Ding, D., Bagul, A., Langlotz, C. and Shpanskaya, K. (2017). CheXNet: radiologist-level pneumonia detection on chest x-rays with deep learning. arXiv preprint arXiv:1711.05225.

Zech, John R., Marcus A. Badgeley, Manway Liu, Anthony B. Costa, Joseph J. Titano and Eric Karl Oermann. (2018). Variable generalization performance of a deep learning model to detect pneumonia in chest radiographs: a cross-sectional study. PLoS Medicine 15(11): e1002683.

8

Tools and Techniques for Disease Prediction

1. Introduction

Artificial Intelligence (AI) tools work with various tools and techniques. The technology advancements help the tools to evolve and better process the data. The tools and innovations in AI that are used for disease prediction systems, have transformed the field of healthcare. The advancements in the tools help to build more efficient and effective early diagnosis tools. This helps in the design of personalized treatment, and accurate evaluations. Disease prediction techniques have helped in reducing the load of physicians as algorithms are used for easing out the decision-making process in the diagnosis and treatment planning. Predictive modeling techniques such as statistical modeling, Machine Learning (ML), and Deep Learning (DL) make use of the advancements in AI techniques. This chapter discusses various tools and techniques for healthcare applications. The chapter also analyses their importance in enhancing the effectiveness and efficiency of the healthcare systems. Tools are designed for processes such as data preprocessing, feature selection techniques, and model evaluation. These are described and other widely used tools and techniques are analyzed. Libraries like Scikit-Learn, Tensorflow, IBM Watson Health, etc. are have also been discussed as a part of this chapter to emphasize their application in solutions for healthcare systems. The different tools helped early researchers to understand various data and use them for available disease prediction. Predictive systems help in identifying and selecting the optimal tools and techniques based on characteristics and problems.

1.1 Purpose of Disease Prediction in Healthcare

Predictive modeling has huge significance in modern healthcare systems. The intelligent system helps to transform patient care with early diagnosis and reduces the risk of patient complications, thus improving survival rates (Obermeyer and Emanuel 2016). Disease prediction models are designed with statistical approaches and AI-ML algorithms. The intelligent systems are capable of providing insights based on

DOI: 10.1201/9781032722344-8

various health data. The algorithms are designed to process data from EHRs, medical imaging, and genomic data. They find hidden patterns and extract relevant features for the study. These intelligent systems assist in disease predictions with diabetes, cardiovascular diseases, and various cancers. The intelligent models applied for clinical decisions help for the identification of high-risk patients. The AI models help for the design and monitoring of personalized treatment and targeted interventions in healthcare systems that monitor and control the prevention of disease progression.

1.2 Significance of Predictive Modeling in Clinical Settings

The integration of predictive models helps clinical workflows. Integration of predictive models helps to process large amount of data and the design of intelligent systems that are data driven. The data driven systems are designed and developed with data from various healthcare applications. Disease prediction models help to improve the diagnostic accuracy and reduce the cognitive load on healthcare professionals. The intelligent assistive tools help to handle the rise in patient requests, when the healthcare system needs to accommodate huge patient volume and associated complexities (Topol 2019). ML-based diagnostic systems can analyze and find patterns from patient data that may not be easily identified by human healthcare practitioners. Application of AI tools to understand the underlying data in healthcare applications could help radiologists to identify abnormal patterns in the imaging data (Rajpurkar et al. 2018).

Predictive models are also designed for the management of chronic diseases. Predictive models are able to identify and assist with the early diagnosis of conditions like diabetes and hypertension. Predictive systems provide continuous monitoring with early alerts and recommendations for treatment and health interventions. These systems assist the healthcare providers to respond and take care of patient health issues well in advance. The intelligent systems can be connected with data from wearable devices to enhance disease predictions and patient monitoring. Predictive models contribute significantly to the quality of care and patient safety by enhancing the precision and reliability of disease prediction.

1.3 Technological Advancements in Disease Prediction

Disease prediction systems are constantly being innovated with technological changes, that keep evolving. The systems are designed to learn and analyze various healthcare data to make intelligent decisions. The systems are designed to make predictions about diseases and progression. Intelligent systems use data and identify the relationships of diseases and associated factors. The evolution of disease prediction systems has made advancements with AI algorithms and technologies. These systems make the AI systems act more precisely, efficiently, and present scalable solutions for predicting and managing health conditions. The advanced disease prediction models are based on large intelligent DL models that are capable of processing large amounts of data. Disease prediction systems traditionally relied on statistical methods, such as logistic regression; recent advancements have helped with the adoption of ML, DL, and data mining techniques. These techniques could analyze large amounts of complex data from EHRs, wearable devices, imaging data, and genomic information (Esteva et al. 2019).

Integrating big data and cloud computing are technologies that have facilitated the design of systems that can process data in real-time. Predictive systems are designed to assist doctors and healthcare practitioners with real-time insights and recommendations. Predictive analytics tools along with AI systems and cloud systems like IBM Watson Health and Google DeepMind, are designed to work for various healthcare tasks. These tools assist with diagnostic support, drug discovery, and personalized treatment. Tools like IBM Watson Health and Google Deepmind are trained on large datasets. The tools assist and work with real-time predictions with high accuracy; making the systems adaptable to provide real-time monitoring and interventions.

Diverse approaches have been applied to predict health outcomes and support clinical decision-making. Traditional statistical models are often limited in handling high-dimensional data and nonlinear relationships. The healthcare systems need the capability to handle large datasets with complex relationships (Goldstein et al. 2014, Goldstein et al. 2017). ML algorithms learn and understand complex patterns from healthcare data. ML models trained on huge amounts of data can mine hidden patterns for accurate prediction (Hamilton et al. 2021). DL algorithms process large amounts of unstructured data. AI systems equipped with DL models effectively process and analyze medical images and clinical notes with minimal intervention (Shickel et al. 2017, Poongodi et al. 2021). AI algorithms are combined together to work as ensemble models; they can effectively handle multidimensional data and map the observations to provide better results in patient care.

It is important to choose the appropriate tools for analysis. Various tools and techniques can be applied to disease prediction. The selection of tools depends on the disease and data type; along with data, the AI system's purpose will also determine the tools needed to design AI systems for optimal performance. For example, logistic regression will work for most binary classification problems, including complex diseases with multiple contributing factors, such as cancer, which may benefit from DL and ensemble approaches that combine numerous AI algorithms (Alanazi et al. 2022). Before selecting appropriate tools and techniques, the data, problem, and need are to be analyzed; this helps in selecting the most suitable methods for their specific applications.

2. Methodologies in Disease Prediction

Various AI technologies are applied for disease prediction systems in healthcare. Intelligent systems are designed to work with AI and computational algorithms and keep evolving into better systems with the use of high computational power, data availability, and algorithmic innovations. Figure 1 shows multiple disease prediction techniques that help design and develop prediction models, covering ML, data mining, predictive analysis, and statistical modeling. The different prediction models that are based on statistical modeling, ML, DL, data mining, and predictive modeling have various advantages and specific applications. The application of computational

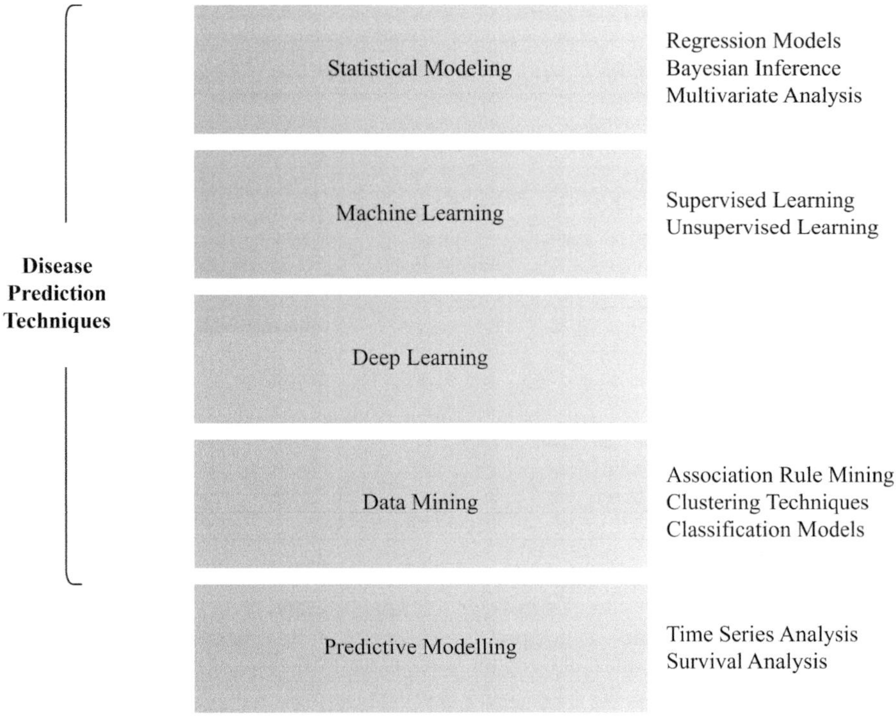

Figure 1. Disease prediction techniques.

methods in disease prediction has made significant progress in healthcare systems. Application of intelligent methodologies are required for the systems to process and analyze complex patient data. Intelligent techniques in the system help to identify complex relationships and patterns and predict potential diseases.

Table 1 compares the various disease prediction techniques with their advantages, limitations, and applications in real-time use cases. The benefits of applying prediction and modeling techniques like (Statistical modeling, ML, DL, Data mining, and Predictive modeling) are discussed in detail to understand the various opportunities with the application of these tools in healthcare systems.

2.1 Machine Learning

ML is an important tool for disease prediction by processing and analyzing complex healthcare datasets. These AI algorithms learn from the patterns and relationships in the data. ML algorithms can be categorized into supervised, unsupervised, and DL methods. Supervised learning work on labeled data for disease classification and early diagnosis makes it an important tool for predicting disease progression. Unsupervised learning is the process of learning from unlabeled data to understand hidden patterns

Table 1. Comparison of disease prediction techniques.

Technique	Description	Advantages	Limitations	Applications
Statistical Modeling	Relies on mathematical equations to predict outcomes	Simple, interpretable, clinically trusted	Assumes data linearity, limited scalability	Risk prediction for chronic diseases
Machine Learning (ML)	Learns patterns from data without explicit programming	Flexible, handles non-linear relationships	Requires substantial data, less interpretable	Imaging and EHR-based disease detection
Deep Learning (DL)	A subset of ML for high-dimensional data like images	Processes unstructured data, high accuracy	Computationally intensive, less transparent	Medical imaging, genomic data analysis
Data Mining	Extracts patterns and relationships from large datasets	Useful for exploratory insights	Limited prediction accuracy	Identifying symptom-disease relationships

from the data. The clustering of similar patient profiles and discovering new disease subtypes is an application of unsupervised learning in healthcare systems. DL is a subset of ML that uses neural networks to process high-dimensional data like medical images, genomic sequences, and sensor data. These ML techniques work for the precise processing of data and help with automated diagnostics. These approaches enhance the accuracy and efficiency of disease prediction, contributing significantly to personalized medicine and improved patient care.

Supervised Learning: Supervised learning is a learning approach in ML; supervised machine learning algorithms need labelled datasets. The algorithms learn from the data; each of the input is associated with a known output. Supervised learning is primarily used for classification tasks. ML algorithms including Decision Trees, Support Vector Machines (SVM), and Random Forests are used to build ML models for healthcare applications. SVM has proven its effectiveness in classifying disease types based on biomarker data (Chouhan et al. 2020). Random Forest model has the ability to handle missing values in medical datasets. Algorithms like Decision Trees, Random Forests, and Support Vector machines are applied in healthcare applications. Decision Trees and Random Forests algorithms are applied for various tasks; the models designed with these algorithms are known for their interpretability. They are effective in handling numerical and categorical data. Decision Trees create a hierarchical model from the data and split the data into classes based on feature values. The Random Forest algorithm combines multiple decision trees; combining multiple decision trees and selecting the one with the better performance helps to enhance the accuracy of Random Forest models (Breiman 2001). Random Forest models are used for the prediction of complex diseases like diabetes and cardiovascular disease. SVM is a robust classification technique used in disease prediction to separate data points by finding the optimal hyperplane. SVM has been applied successfully to classify tumor malignancy in oncology.

Unsupervised Learning: The unsupervised learning technique works with unlabeled data. In unsupervised learning, the learning process happens from the data by identifying patterns and forming clusters. It is ideal for discovering hidden patterns from large datasets. Clustering techniques such as k-means and hierarchical clustering are applied to group patients based on similar characteristics (Jiang et al. 2017). Unsupervised learning processes unlabeled data to find hidden patterns. Clustering techniques are applied to form clusters of identical patient profiles for discovering new disease subtypes. It is also applied for anomaly detection to identify unusual patterns that may indicate rare disease outliers in patient data. Commonly used unsupervised learning techniques include algorithms that group similar data points together which can help in identifying patient subgroups with identical risk profiles. Anomaly detection is used to detect rare disease cases and abnormal health conditions in populations in the unsupervised learning techniques.

Deep Learning (DL): DL is a subset of ML that processes a high volume of multi-dimensional data. DL algorithms are applied to process medical imaging and genomic data (Esteva et al. 2019). Convolutional Neural Networks (CNNs) is a DL technique that effectively processes medical images. CNN-based models are more efficient in predicting and identifying various diseases and regions of interest than expert radiologists (Rajpurkar et al. 2018). Recurrent Neural Networks (RNNs) and Long Short-Term Memory (LSTM) networks are used in the analysis of sequential data such as patient health records. RNN and LSTM are used with the time series-based models for data analysis for understanding disease features and building systems that can forecast disease impacts (Shickel et al. 2017, Poongodi et al. 2021).

2.2 Data Mining Techniques

Data mining techniques are used in disease prediction systems to identify and extract patterns, associations, and trends from datasets. Data mining techniques process complex clinical data to identify relationships. Identifying relationships and selecting trends and patterns from data detects the correlations between symptoms, risk factors, and diseases. The patterns from multimodal patient data give insight into hidden patterns that are cannot be observed easily and facilitates accurate disease prediction. Pattern mining contributes to a better understanding of disease mechanisms. Data mining helps with early diagnosis, personalized treatments, and improved patient care. Classification, clustering, and association are data mining techniques that are used to identify rules and patterns from data. Data mining is used by healthcare professionals for data-driven decisions. They improve the overall effectiveness of medical interventions.

Association Rule Mining: Association rule mining is identifying the relationships between variables from large databases. Apriori and FP-Growth are association rule mining algorithms that work based on determining the frequency of item sets for generating rules. This helps identify and highlight common patterns and helps identify symptom clusters from patient records. Association rule mining identifies and categorizes risk factors and disease parameters by analyzing patterns extracted from data.

Clustering Techniques: The clustering technique is used to group similar data together. Clustering helps with the identification of subpopulations based on patient data. K-means clustering, groups patients based on the disease severity and associated risk factors. The grouping of patients based on severity helps to give targeted treatments based on their requirements. Hierarchical clustering, groups diseases based on shared causes and symptoms, contributing to the advancement of personalized medicine.

Classification Models: Classification models are used in healthcare systems for the prediction of disease indicators. The classification methods work better with binary and multiclass prediction applications. Classification models like logistic regression, Naïve Bayes, and Decision Trees help with the prediction of diabetes and heart disease (Topol 2019). Logistic regression models have high interpretability of the models; Decision Trees and ensemble methods like Gradient Boosting perform better for handling complex datasets and with higher accuracy and flexibility.

2.3 Predictive Modelling

Predictive analysis works with statistical and ML techniques. They are used to process historical and mealtime data for the forecast disease patterns and outcomes. Predictive analysis is used by healthcare professionals to make informed decisions about patient care. This approach is practical in managing chronic diseases and for the early detection of risk factors and disease progression. The early detection of the diseases will help to improve treatment efficiency. Predictive analysis helps for the personalized treatment approaches that help with the design of preventive healthcare systems. Intelligent systems have the ability to forecast potential health issues. The predictive systems are used for the timely intervention and the design of personalized healthcare strategies to reduce the adverse impacts of diseases on individuals.

Time series analysis allows healthcare data to be processed with a temporal dimension. The real-time data captured using the sensors are monitored and processed with the help of time series analysis algorithms. Time series analysis and its applications ensure the real-time monitoring of disease progression and detect anomalies in patient vitals. Autoregressive Integrated Moving Average (ARIMA) models are used to analyze time-dependent health factors like glucose levels in diabetic patients (Kaushik et al. 2020). Time series data analysis has the role and significance with designing systems in healthcare applications that can constantly monitor patients.

Survival analysis is a specialized predictive analytics approach designed to estimate the time until a specific event of interest. This includes the case of disease progression or patient mortality. Kaplan-Meier estimators and Cox proportional hazards models are used for cancer prognosis and medical predictions (Bonato et al. 2011). These models provide insights into survival risk factors. The survival analysis and identifying the risk factors helps healthcare providers to identify high-risk individuals and develop personalized treatment strategies. Time-to-event data survival analysis enhances decision-making in clinical practice and supports better management of time-sensitive health conditions.

2.4 Statistical Modelling

Statistical models are one of the primary tools that are used in disease prediction systems. Analyzing and processing complex data enables healthcare process data and find relationships. These models provide a framework with rigorous hypothesis testing. Hypothesis testing allows the evaluation of relationships between various health parameters and disease features. Statistical models work with a probabilistic explanation of data. Using techniques like regression analysis, survival models, and Bayesian inference, intelligent systems learn from the data and predict the future. Statistical modeling helps with the identification of risk factors and predicts the likelihood of disease occurrence and progression. Statistical techniques enable users to get deeper insights from patient data. These methods work on data, enhance the accuracy of diagnostic predictions, and develop evidence-based approaches for disease prevention, treatment, and management.

Regression analysis is an important statistical tool that can be used for disease prediction. Regression analysis provides a method for quantifying relationships between variables. Linear and logistic regression models are used in healthcare applications to predict outcomes based on known risk factors (Goldstein et al. 2017). Logistic regression solves problems related to binary classification tasks, including determining a particular disease's presence or absence. Bayesian models are used to incorporate prior knowledge about the condition before making predictions. Bayesian models are suitable for medical applications with limited data. Bayesian networks have been applied in scenarios like predicting patient treatment responses and modeling the probabilistic relationships between disease symptoms (Spiegelhalter et al. 2004). Multivariate analysis considers multiple factors simultaneously while processing before making predictions. This helps them to analyze and make use of the complex interactions between variables. Methods such as factor analysis and PCA are used to reduce data dimensionality. These methods are also efficient in identifying hidden patterns from data. Multivariate analysis is essential for the examination of multiple biomarkers and genetic factors that contribute to disease risk predictions (Van der Maaten and Hinton 2008).

3. Feature Selection for Disease Prediction Models

Feature selection process is essential in identifying and selecting the best attributes that are essential in developing predictive models. The feature selection technique identifies the best attributes based on the disease type the model should work with. Identifying the best features is vital in disease prediction systems and while designing models to make the predictions. The feature selection and the identified features are directly related to the model performance, interpretability, and computational efficiency. Identifying the most relevant features from the dataset is essential. Feature selection methods are used to reduce the dimensionality of the data. The directionality reduction process does not remove the essential information. Selecting the best techniques improves the predictive accuracy and generalizability (Chandrashekar and Sahin 2014). In healthcare systems, data often involves numerous clinical, demographic, and biological variables. Effective feature selection process helps to improve disease prediction. The process focuses on selecting the key features from the data that are required for the disease risk and progression.

3.1 Importance of Feature Selection

Feature selection is an important factor in building healthcare models and AI systems based on the most important attributes. The feature selection process solves several challenges with healthcare data. The high dimensional and complexity issues can be reduced with intelligent selection of the best features. Feature selection technique reduces the model complexity. Feature selection also solves the issues with model training. Issues like overfitting is a common problem when working with irrelevant variables to build a predictive model (Guyon and Elisseeff 2003). Feature selection reduces the complexity of models and enhances its interpretability allowing clinicians to identify which factors contribute most significantly to disease risk. This is particularly important when models are built for precise medicine where understanding the role of specific biomarkers or lifestyle factors can inform tailored interventions (Kohavi and John 1997). Feature selection technique removes irrelevant features after careful consideration, improves computational efficiency, and allows faster processing and analysis on large datasets.

3.2 Methods for Feature Selection

Feature selection techniques are categorized into three main types: filter, wrapper, and embedded methods; each applies distinct approaches to identify and select relevant features from a dataset. Filter methods rely on statistical measures to rank features based on their importance and choose the most significant ones. Filter methods are computationally efficient models. Wrapper methods evaluate subsets of features by training and testing a model. Wrapper models provide higher accuracy but have increased computational complexity. Embedded methods integrate feature selection directly into the model training process, balancing efficiency and performance. The choice of the best technique depends on factors like the size of the dataset, feature dimensions, and the objectives of the predictive model. The feature selection technique identifies and selects the optimal feature selection to enhance model accuracy and interpretability.

3.2.1 Filter Methods

Filtering methods are feature selection techniques based on statistical values. Filtering techniques rank the features based on their relationship to the target variable. Commonly used feature selection techniques by the filtering method are correlation-based and mutual information-based selection. The correlation-based filtering technique identifies and selects the features based on the correlation value between features and the target variable. In mutual information, the amount of information each feature provides about the outcome is considered.

Correlation-based feature selection is often used to predict cardiovascular disease, in which age, cholesterol levels, and blood pressure significantly correlate with the target variable disease risk (Hamilton et al. 2021). Filter methods are computationally efficient and the technique looks at the feature interactions. Application of filtering technique for complex data sets is challenging.

3.2.2 Wrapper Methods

Wrapper methods are used to select the features by evaluating the feature subsets by training and evaluating a predictive model on different feature combinations. Standard techniques used for wrapper-based feature selection include recursive feature elimination (RFE) and forward/backward selection. In forward/backward feature selection, iterative addition and removal of features are done to identify the best subsets that help to optimize model performance (Kohavi and John 1997). Wrapper methods are computationally intensive; they often yield more accurate results than filter methods due to their ability to account for feature interactions. Wrapper methods have succeeded in disease prediction tasks involving complex, multivariate datasets, such as predicting diabetes risk using clinical and lifestyle data (Hall 1999).

3.2.3 Embedded Methods

Embedded feature selection techniques integrate and process the feature selection strategies. This technique works directly during the model training process. The model automatically prioritizes features based on their predictive contribution to the embedded technique. Techniques such as Lasso and Ridge regression apply regularization penalties to minimize the influence of less significant features, effectively selecting only the most relevant ones (Tibshirani 1996). ML algorithms based on tree architecture, the Decision Tree, and the Random Forest algorithm include built-in mechanisms to evaluate and rank feature importance during model building. Embedded feature selection method is computationally efficient and effective in selecting and mapping the importance of features. The embedded feature selection method performs the feature selection process as part of the modeling process by eliminating the need for a separate feature selection process. Lasso regression is commonly applied in genomic data analysis for disease prediction, effectively identifying key genetic markers while reducing irrelevant and redundant features. This ability of identifying the key features by removing the redundant and irrelevant features makes lasso regression ideal for handling high-dimensional datasets (Zou and Hastie 2005).

3.3 Impact on Model Performance

An effective feature selection process is required to improve the predictive models; models learn from the data and the clean data helps the algorithms to identify the best patterns and learn. One of the main aims of feature selection is to select the most appropriate features for the modeling. Another advantage while doing the feature selection is that the feature selection process reduces the noise in the data and focuses on selecting the most informative features. This process improves model accuracy, reduces overfitting, and improves model generalizability (Guyon and Elisseeff 2003). Identifying and selecting the right features is crucial for modeling and processing huge datasets. Identifying genetic mutations and protein expression levels in cancer prediction models needs the creation of the best models that learn from complex data. These models must select and learn from the best features that can give high accuracy, better performance, and provide insights into the disease mechanisms (Sajjadnia et al. 2020). The feature selection process allows the development of interpretable models. These models help to explain the significance of each of the features. This helps improve models and increases the practical applications and clinical trust of AI models.

The feature selection process impacts the computational demands and raises the complexity of developing predictive models. High-dimensional datasets are an essential component of healthcare applications. Applications such as genomics and medical image processing require huge processing power and memory to handle the design and development of AI models for predictive applications. Reducing the number of features is a flawed approach to these modeling requirements. The feature selection techniques can be applied to identify the best features and increase computational complexity. The huge requirement for computation data storage makes the models challenging for real-time applications (Liu and Motoda 2012). Healthcare systems have adopted AI and ML tools to improve patient care. Efficient feature selection and modelling approaches are required to build responsible and scalable predictive models.

3.4 Case Studies on Feature Selection in Disease Prediction

Various studies have shown the importance of having feature selection when working with a large dataset. This helps with the improvement of performance of intelligent systems for disease prediction and forecast. For a model that is designed for heart disease prediction, incorporating filter-based feature selection has improved model accuracy by focusing on key risk factors such as age, blood pressure, and cholesterol levels (Topol 2019). A study on diabetes prediction using the wrapper feature selection method has demonstrated improvement in model performance. The model performance was improved by modeling using a subset of features that are most relevant to diabetes risk (body mass index (BMI) and fasting blood sugar levels) (Goldstein et al. 2017). These case studies demonstrated that feature selection has a better impact when the most important features are used for building AI models. Identifying the best features from the dataset enhances the explainability of the model and improves its clinical acceptance.

Embedded feature selection methods using Lasso regression have shown to be effective in genetic research for disease prediction. The application of Lasso regression has helped to isolate significant genetic variants from a large genomic data pool. This has led to breakthroughs in predicting diseases such as cancer and

Alzheimer's disease based on genetic predisposition (Zou and Hastie 2005). These studies highlight how feature selection enhances the predictive accuracy of the AI models and contributes to a deeper understanding of disease mechanisms.

4. Data Preprocessing in Disease Prediction

Data preprocessing is an important step when developing predictive models. The data for healthcare applications requires cleaning and fine-tuning before being used for modeling and AI applications. The raw data needs to be carefully prepared for the analysis. The data needs to be clean, consistent, and suitable for the chosen analytical operation. Healthcare data is complex and combines data from sources, including clinical records, genetic data, medical imaging, and sensor data. Each of these data has its unique features and challenges. Data preprocessing techniques are required to improve the quality of data. Preprocessing steps helps with the improvement of data quality. Good quality data, when used for modelling improves the model accuracy, reduces noise, and ensures reliable predictions. Key data preprocessing techniques help to remove the challenges with healthcare datasets, by handling missing data, normalization, outlier management, and dimensionality reduction (Cabello-Solorzano et al. 2023).

Figure 2 illustrates the key preprocessing techniques, which include handling missing data, normalization, outlier detection, and dimensionality reduction. These methods are essential for preparing raw data to ensure it is clean, consistent, and suitable for analysis. Table 2 summarizes the preprocessing techniques and their benefits and challenges.

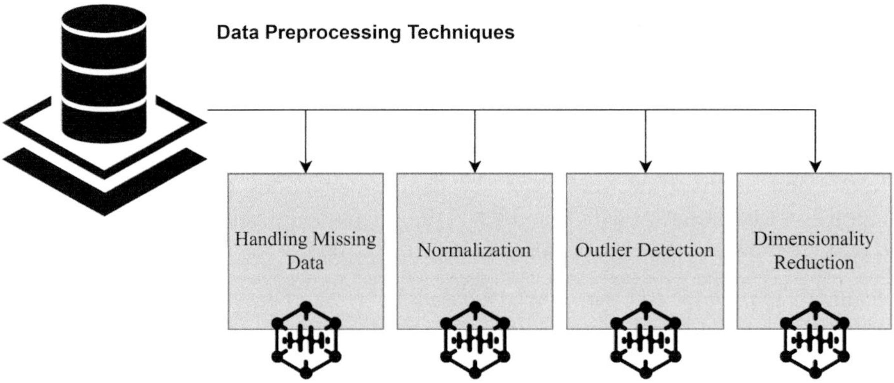

Figure 2. Data preprocessing techniques.

Table 2. Data preprocessing techniques.

Preprocessing Technique	**Description**	**Benefits**	**Limitations**
Handling Missing Data	Filling or removing incomplete data	Reduces bias, improves accuracy	May introduce assumptions if not handled carefully
Normalization	Scaling data values to a standard range	Enhances algorithm performance	May lose the interpretability of raw values
Outlier Detection	Identifying and managing extreme values	Removes noise, improves model reliability	Risk of losing important rare disease indicators
Dimensionality Reduction	Reducing feature space (e.g., PCA, t-SNE)	Improves computational efficiency	May discard relevant features

4.1 Handling Missing Data

Missing data is a frequent issue with healthcare data. The issue is due to incomplete records on patients, this can be due to patient noncompliance or errors during the data collection process. Handling missing values is one of the major steps in making data consistent. If missing values are present in the data and not handled properly, it can lead to biases in the model performance and have to be addressed carefully (Little and Rubin 2019). Techniques for handling missing data vary depending on the amount and type of missingness in the data. Missing data can be categorized as Missing Completely at Random (MCAR), Missing at Random (MAR), or Missing Not at Random (MNAR). In order to handle missing values imputation methods and deletion of data are applied by careful examination of data.

Imputation Methods: Data imputation methods are simple methods that fill the missing columns with values like mean, median, or mode values. The imputation method usually uses the representative values from the dataset. For the best imputation to fill the missing values, ML-based methods like k-nearest neighbors (KNN) and multivariate imputation by chained equations (MICE) are used. In imputation techniques of MICE and KNN the relationship from the data is analyzed to provide the best values to be filled (Azur et al. 2011).

Deletion Techniques: In cases where the data have major features missing, deletion (removing rows with missing values) can be used. This is a destructive way of preprocessing, where the data is removed from the analysis or modeling; it is required to accommodate the model's effectiveness to remove the brassiness. Pairwise deletion is another alternative, where only complete pairs of data points are used for correlation calculations, preserving more of the dataset (Little and Rubin 2019).

Proper handling of missing data is critical in disease prediction. Error handling techniques ensure that the model predictions are unbiased. Failing to handle the missing data in patient demographics or clinical markers in cancer prediction models can skew risk estimates (Li et al. 2017).

4.2 Normalization and Scaling

Normalization and scaling are important preprocessing steps that help to standardize data. These techniques ensure that the data range is scaled and ensures that all variables contribute proportionally during model training. Normalization and scaling techniques are important in healthcare applications. The data comes from various sources, this includes test results, patient age, and vital signs. This data needs proper scaling of values. By bringing these features to a comparable range, normalization and scaling improve the performance and stability of predictive models, allowing them to capture relationships within the data better.

Normalization techniques scale the values to a standard range. Min-max scaling is a data normalization technique to re-scale the values to a specific range (0 to 1). These techniques remove bias from data when working with distance-based ML models. The differences in scale could affect the performance of ML models (Kuhn 2013). Data standardization using the Z-score normalization technique helps transform features into a mean of zero and a standard deviation of one. This technique benefits algorithms like logistic regression and support vector machines, which usually assume distributed data (García et al. 2015).

Applying these data transformations significantly improves the quality of data. Quality data helps improve model performance. Data transformations play a significant role in developing predictive models involving EHRs, imaging data, or genomic data, where variations in feature scales are common.

4.3 Outlier Detection and Management

Outliers are extreme observations or values in specific observations that deviate from the common domain values for an observation. Outlier values impact the performance of AI models. The data will get skewed when outlier values are present. Outliers in the data indicate measurement errors, rare conditions, or unique patient characteristics. Careful observation has to be made to handle outliers. Detecting and managing outliers is one of the main steps to ensure the reliability of predictive models.

Statistical techniques like Z-score, IQR (Interquartile Range), and Tukey's fences detect outliers by identifying data points which are present outside a defined range. In clinical data, Z-score-based detection is widely applied to identify values that deviate significantly from the mean. Intelligent algorithms such as Isolation Forests and Local Outlier Factor (LOF) are practical for outlier detection, particularly in complex datasets with insufficient traditional statistical methods. Isolation Forests process EHR data and detect abnormal lab results and patient history patterns. The abnormal results and observations highlight data errors or rare diseases (Liu et al. 2008).

Effectively managing outliers is essential for disease prediction models. When studying chronic diseases, extreme values in health metrics (e.g., blood glucose levels) may represent disease representations rather than noise.

4.4 Dimensionality Reduction

Healthcare datasets are data from multiple sources and are usually high-dimensional. The data from genomics and medical imaging sources are generally represented as big data, with data from various dimensions. High-dimensional data have challenges with training issues in models. There is an increase in the risk of overfitting and interpretability issues of results. Dimensionality reduction techniques, such as Principal Component Analysis (PCA) and t-distributed Stochastic Neighbor Embedding (t-SNE), address the problems with big data and the presence of considerable dimensions in data. The dimensionality reduction techniques select the most critical features and retain essential information for the AI algorithms to learn the patterns.

PCA is a linear dimensionality reduction method that transforms features into principal components. PCA is used for disease prediction and to reduce high-dimensional data complexities. In genetics and brain imaging studies, PCA reduces the large data volume and helps concentrate on significant factors, making processing easier (Wold et al. 1987). t-SNE is a non-linear technique for visualizing high-dimensional data by mapping data points into a two-dimensional or three-dimensional space. The process is computationally intensive but effectively identifies clusters from patient data. This process is helpful to identify and cluster disease subtypes for treatment planning (Van der Maaten and Hinton 2008).

Dimensionality reduction is a process that is applied to simplify data and enhance model performance by reducing noise. Processing and cleaning data requires focusing on the most informative features. When working with complex disease studies, selecting the best features that represent the data is needed. When predicting neurological disorders or other significant diseases, high-dimensional data must include all possible causes.

4.5 Case Studies on Preprocessing Effects

Researchers have completed several studies emphasizing the need for clean and proper data. The impact of data preprocessing was evident from various research studies that identified effective preprocessing techniques. Clean data has a direct effect in improving the accuracy of prediction models. For example, in a study that focused on diabetes prediction, researchers used normalization techniques to normalize the data and impute missing values. These data processing techniques have enhanced the model performance for identifying high-risk patients (Topol 2019). Such preprocessing steps help to process and solve the issues with inconsistencies and missing data from the dataset. Missing and unbalanced data leads to skewness in the results for the biased prediction.

Another study on breast cancer prediction highlighted the need for dimensionality reduction techniques. The PCA technique helped to identify the essential features of the data. Extracting the crucial features using PCA has improved the model's accuracy and enhanced interpretability by reducing redundant and irrelevant features. Identifying the principal features makes it easier for clinicians and researchers to understand the underlying patterns and relationships from the data (Li et al. 2017).

The feature selections and reducing data dimension and complexity have contributed to faster computation and a more efficient use of computation resources. Effective use of research and lightweight models is essential when creating models with high-dimensional datasets in healthcare.

These examples demonstrate that robust data preprocessing is far more than a preparatory step; it is a critical foundation for building effective and reliable predictive models. By improving data quality, consistency, and relevance, preprocessing ensures that models can achieve their full potential, delivering accurate predictions and actionable insights. Data preprocessing and cleaning are essential for cleaning the data for the better representation of data and its features. Each case requires customized preprocessing strategies as each data varies with the data and its features. Working for specific characteristics and challenges of the dataset is needed to enhance the ability of the model to support data-driven decision-making in healthcare.

5. Model Evaluation Techniques for Disease Prediction

AI models are developed to assist patients and healthcare systems to ensure intelligent decision-making. Evaluating the performance of the predictive models is crucial with analyzing the accuracy of these models. Evaluation techniques help to ensure that the models are reliable for the disease prediction systems. Predictive outcomes are directly influenced by patient care and treatment plans. There is a need for building robust and accurate models for healthcare applications. Evaluation techniques assess and ensure effective model performance and reliability. Several key factors affect model performance evaluation; the evaluation metrics are designed to capture and analyze model performance. Accuracy, confusion matrix, cross-validation, etc., are some strategies that help evaluate models. The evaluation metrics also help identify and select the best models for clinical implementation.

5.1 Evaluation Metrics

Selecting the best evaluation metrics is essential for assessing model performance. In order to evaluate the AI models there are various evaluation metrics. Each of the evaluation metrics is designed to capture the distinct aspects of the model created. This includes parameters like accuracy, reliability, and clinical relevance. Standard evaluation metrics that are used for evaluation of the models include accuracy, precision, recall, F1 score, confusion matrix and area under the receiver operating characteristic curve (AUC-ROC).

Accuracy: Accuracy measures the proportion of correct predictions among all predictions made by the model. The accuracy metric is simple and easy to understand but the accuracy alone cannot be relayed as an effective evaluation metric. This metric can be misleading when modeling with an imbalanced dataset where observations belonging to one class are more than the other (Chicco and Jurman 2020). In cancer screening models, high accuracy may be achieved by predicting the majority class (no cancer) without capturing actual positive cases. Thus, the accuracy metric alone cannot be used to evaluate the performance of the AI models in healthcare applications.

Precision and Recall: Precision indicates the proportion of true positive predictions among all positive predictions made by the model. Recall, also known as sensitivity, measures the proportion of actual positive cases the model correctly identifies. When designing models for prediction applications, it is required to ensure that the models have high precision so that false positives are minimized and high recall so that the models can capture true disease cases, preventing missed diagnoses (Sokolova et al. 2006, Powers 2020). Precision and recall evaluation metrics help with the evaluation of the models that are used for prediction applications that can handle severe diseases where false negatives (missed diagnoses) have serious consequences.

F1 Score: The F1 score is the harmonic mean of precision and recall, providing a balanced evaluation metric that is especially useful in imbalanced datasets. F1 score combines false positive and false negative, making it a suitable evaluation metrics for healthcare applications, where both errors and accurate predictions have significant roles. The F1 score is often used for evaluating the diagnostic models where a balance is achieved between precision and recall (Christen 2023).

AUC-ROC and PR Curves: The AUC-ROC and Precision-Recall (PR) curves are used to evaluate model performance across different threshold values. The curves represent the trade-off between sensitivity and specificity. PR curve shows the precision-recall balance. AUC-ROC and PR curves evaluation metrics is helpful when a model is being trained using an imbalanced dataset (Davis and Goadrich 2006). Disease prediction models with high AUC-ROC score means that the model can identify positive and negative cases effectively (Bradley 1997).

Confusion Matrix: Confusion matrix represents model predictions and classifies the predictions into true positive, false positive, true negative, and false negative predictions. Confusion matrix offers detailed insights into the model performance. It is particularly useful in healthcare applications as it identifies specific error types (e.g., false negatives) and helps to form strategies for improvements for the model (Tharwat 2021).

Table 3 summarizes the key evaluation metrics commonly used for predictive model assessment and their definitions, applications, and associated challenges.

Table 3. Evaluation metrics for disease prediction models.

Metric	**Definition**	**Use Case**	**Limitations**
Accuracy	Proportion of correct predictions	General-purpose	Misleading in imbalanced datasets
Precision	True positives out of all predicted positives	Reduces false alarms	Low recall could miss actual cases
Recall (Sensitivity)	True positives out of actual positives	Critical for disease detection	Low precision increases false positives
F1 Score	Harmonic mean of precision and recall	Balance between recall and precision	Requires both metrics to be calculated
AUC-ROC	Area under the sensitivity-specificity curve	Evaluates model's discrimination ability	May not account for dataset imbalance

These model performance evaluation metrics play a significant role in analyzing the performance, reliability and efficiency of the model. In healthcare applications the models need to be accurate and the predictions made by these models are critical. The table highlights specific use cases for each evaluation metric and the limitations to consider when applying them.

5.2 Cross-Validation Techniques

Cross-validation is tracked and ensures the generalizability of ML models and this ensures that the ML model is performed unbiased on various sample populations. ML Models for healthcare applications are to be unbiased and tested for the model performance with various patient profiles. Cross validation is applied by partitioning the dataset into multiple subsets. Cross-validation techniques estimate the model performance with unseen data and reduce the overfitting that can be caused while training the model. The main cross validation techniques are K-Fold Cross-Validation, Leave-One-Out Cross-Validation (LOOCV) and Stratified Cross-Validation.

K-Fold Cross-Validation: K-fold cross-validation technique splits the dataset into k equal parts. The model is trained with k-1 folds and validated on the remaining fold. This training and validation process is repeated for k times and the process of training and validation is continued till each fold has been served as the validation set once. The results are averaged to produce a reliable performance estimate model (Hastie et al. 2017). K-fold cross-validation technique is used to validate and estimate the performance of disease prediction models. K-fold cross-validation provides a strong measure of model performance across multiple subsets of the data.

Leave-One-Out Cross-Validation (LOOCV): LOOCV is an extreme form of K-fold cross-validation. In LOOCV technique k is the number of data points in the dataset. In LOOCV each data point will be used as a validation set once. This process is computationally expensive and is highly effective for small datasets. This technique maximizes the ability of the model to be trained on the data. The patterns from the data are effectively captured in this technique and are helpful when working with rare disease studies and limited patient data (Jung 2014).

Stratified Cross-Validation: Stratified cross-validation ensures that each fold maintains the same class distribution as the original dataset. This is especially valuable in imbalanced healthcare data like rare disease prediction. This technique provides more accurate and representative performance estimates for imbalanced datasets, enhancing model reliability in real-world applications (Kohavi 1995).

Cross-validation techniques are crucial to evaluate the model's ability to generalize, particularly in healthcare. Predictive models must perform consistently across diverse populations and clinical environments. These techniques partition the dataset into training and testing subsets, ensuring that the model is trained on one portion while being validated on another. Cross-validation ensures that the data has a balanced distribution of data classes, reducing the risk of overfitting and improving reliability. In healthcare applications cross-validation helps with the assessment of model robustness and ensures its effectiveness in real-world scenarios supporting accurate and reliable predictions for diverse patient groups.

5.3 Impact of Evaluation on Model Selection

Model evaluation techniques are used to understand the performance of AI models. Proper model evaluation techniques are critical in selecting and developing predictive models for healthcare applications. Evaluation metrics help to indicate the performance of the AI models on disease prediction applications. Accuracy and reliability are crucial in the models that are designed for healthcare applications. Models with high recall are prioritized for cancer prediction to reduce the cases. Precision is another evaluation metric that is checked in chronic disease management models. Minimizing false positives is essential to avoid unnecessary interventions (Topol 2019). Each of the healthcare models are tested and evaluated based on various evaluation metrics.

Model evaluation also plays a crucial role in validating the clinical applicability of disease prediction models. A highly accurate model in its experimental settings may not perform as well as a generalized model in a clinical application. The evaluation metrics help to evaluate the performance of the models based on various data points. This ensures that the models are performing well with diverse patient populations. Making sure that the models are effective is particularly relevant in healthcare systems. The various models that are designed must work under varying conditions, demographic characteristics and medical histories (Nayak et al. 2016, Hamilton et al. 2021). The evaluation metrics that consider clinical priorities, such as the F1 score for imbalanced datasets or AUC-ROC for screening sensitivity, help to ensure that predictive models meet the practical needs of healthcare providers and patients.

5.4 Examples of Model Evaluation in Disease Prediction

AI models require rigorous model evaluation especially when used in clinical applications. Model evaluation plays an important role in developing reliable and effective disease prediction models. Proper evaluation helps the models to be accurate and properly assist and meet clinical requirements and expectations. A model prepared for diabetes prediction used the evaluation metrics AUC-ROC and F1 scores to assess the model performance. This model is created to distinguish between high-risk individuals and low-risk ones. The AUC-ROC metric is used to evaluate the model's ability to balance sensitivity and specificity. The F1 score identifies the trade-off between precision and recall. These comprehensive evaluation metrics are used to ensure that the model is capable of accurately identifying high-risk patients without generating an excessive number of false alarms. The evaluation metrics help to validate that the models are not raising wrong predictions and validate the model performance. When applied in healthcare systems ensures they can be used effectively for resource allocation in healthcare solutions (Goldstein et al. 2014, Goldstein et al. 2017).

Another study, aimed to create a model for breast cancer prediction, highlighted the challenges based on imbalanced datasets. The model was trained on the data where one class is more than the other (normal cases are more than the cancer cases). The researchers use stratified K-fold cross-validation to get a balanced representation of classes during model training and testing. PR curves are used to evaluate the ability of the model to detect cancer cases effectively addressing the issues with class imbalance (Sajjadnia et al. 2020). Model evaluation metrics are used to evaluate and enhance the robustness and accuracy of the model. This ensures that the models are fine-tuned and are efficient to be applied for real-world scenarios. There are many case studies emphasizing the need for models that align evaluation metrics and techniques with the specific clinical objectives of disease prediction models; the different healthcare applications often require appropriate approaches to ensure their practical applicability and reliability in diverse clinical settings.

6. Comparative Analysis of Tools and Techniques

Developing predictive models requires the use of the latest tools and techniques. Advanced tools are designed for disease diagnosis and prognosis and require for various computational tools and methodologies. When using the advanced models, the requirement for computational power grows exponentially. With the increased complexity of healthcare data, choosing the appropriate tools and techniques for a specific disease prediction task is also a crucial factor for achieving high accuracy, interpretability, and clinical applicability. This section compares widely used machine learning (ML) and statistical tools in disease prediction. The tools, techniques and their strengths and weaknesses have to be analyzed when selecting the right tools to process various datasets.

6.1 Tool Comparison Matrix

Various software tools and platforms are designed for the development of disease prediction models. Popular python libraries are used for the design of software tools and models. Python libraries such as Scikit-Learn, TensorFlow, Keras, and IBM Watson Health (Figure 3) are used for the development of intelligent tools (The application of these tools are mention in Table 4). These tools are designed to provide various functionalities from basic statistical modeling to complex DL architectures. These open-source python libraries and tools enable researchers and clinicians to prepare customized models to meet their specific needs.

Each of the tools discussed are designed for specific purposes. The design of tools needs to be based on the requirement for the tools. The intelligent systems are designed to solve specific problems for particular types of diseases. Scikit-Learn is well-suited for rapid prototyping and simpler ML tasks. TensorFlow and Keras are used for developing and deploying complex DL models that can process high-dimensional and multimodal medical data. IBM Watson Health provides advanced analytics and is applied to various clinical applications. This tool is suitable for building healthcare systems with less technical flexibility for custom model building (Abadi et al. 2016, Pedregosa et al. 2011).

Figure 3. Tools and packages.

Table 4. Comparison of various tools and packages.

Tool	Key Features	Advantages	Limitations
Scikit-Learn	Offers a comprehensive set of ML algorithms, including regression, classification, clustering, and preprocessing	User-friendly, open-source, extensive documentation, suitable for beginners	Limited support for deep learning
TensorFlow	Supports deep learning and neural network architectures, compatible with high-dimensional data	Scalable, high performance, supports complex neural networks	Has a steep learning curve, computationally intensive
Keras	High-level API for neural networks, built on top of TensorFlow	Simplifies deep learning model building, flexible, widely adopted in academia	Limited control over model internals
IBM Watson Health	Advanced analytics, AI-enabled insights, tailored solutions for healthcare	Cloud-based access to pre-trained models suited for clinical applications	Proprietary may have limited customization options

6.2 Strengths and Limitations of Techniques

Disease prediction systems are designed for the improvement of the performance of AI models. These AI models are trained to be used for advanced patient care systems. The patient care systems use advanced AI models to provide effective patient care. In order to train AI models, a large database is required and the models learn from them and understand patterns from the data. Modern healthcare systems use models that can balance the complexity, data characteristics, and interpretability. The commonly used predictive techniques include statistical modeling, ML, data mining, and deep learning. The strengths and limitations of the various disease prediction techniques are discussed below.

Statistical Modeling: Includes models like logistic regression and Bayesian inference. These approaches are designed to create disease prediction systems. The outputs from these systems can be easily interpreted and simplified to understand. These tools are essential in clinical applications where decision interpretability is critical (Goldstein et al. 2014, Goldstein et al. 2017). Statistical models are limited by their assumptions about data linearity and distribution that restricts their effectiveness with high-dimensional or highly variable healthcare data.

Machine Learning (ML): ML algorithms like Random Forest and Support Vector Machine (SVM) are combined and applied for the design of ensemble methods that have flexibility, reliability and robustness. These models work by the identification of non-linear relationships from the data. Random Forest models are well-suited for disease prediction tasks that have many features that contain complex interactions. The random forest model has the potential to reduce overfitting and focuses on identifying key features of importance (Breiman 2001). ML models are computationally intensive and their lack of interpretability affects the acceptance of applications designed in clinical use cases.

Data Mining: Data mining methods are used for identifying association rules and clustering. Association rules and clustering are instrumental in identifying the hidden patterns from large datasets. They help in easier identification of symptoms and disease associations. Data mining techniques help in exploratory analyses and identifying risk factors. Association rule mining helps to detect hidden patterns from the data (Li et al. 2017). Data mining techniques lack predictive accuracy and require additional methods to turn the insights into actionable disease predictions.

Deep Learning: DL models like the Convolutional Neural Networks (CNNs) and Recurrent Neural Networks (RNNs) are effective to handle unstructured data such as medical imaging and time-series data. CNNs are effective in identifying and segmenting tasks for tumor detection in radiology. CNNs are effective for identifying minute patterns from imaging data that may not be visible through human eyes (Litjens et al. 2017). DL requires extensive computational resources and large datasets to achieve high accuracy, and the black-box of DL algorithms raises challenges for interpretability and clinical trust.

6.3 Matching Tools and Techniques to Disease Types

The choice of tools and AI techniques for building the models need to be associated with the applications. The nature of the disease and the data type are some of the important factors that play an important role in deciding the tools. Diseases with a genetic basis, such as cancer, benefit from DL models that analyze high-dimensional genomic data. For diseases where symptom patterns are crucial like diabetes, combining various data mining techniques (for feature selection) and ML models (for classification) may offer the best and reliable results.

Chronic Diseases: Chronic diseases including diabetes and hypertension need the analysis of time-series data. These diseases progress and vary over time. The analysis of the data helps to understand the disease and its patterns in the patient. Techniques like time series analysis and ML models are effective with the processing of real-time time-series data. These techniques can help with the identification of the trends and relationships from longitudinal data (Zeger et al. 2006). Python Libraries including Scikit-Learn and TensorFlow help with the analysis of the data for chronic diseases.

The AI models are designed to handle various heterogeneous data along with model building and integration with real-time time-series data.

Infectious Diseases: Infectious diseases including COVID-19 and Influenza require the modelling of AI tools that can provide rapid and accurate predictions. Logistic regression and decision tree-based models are suitable to get interpretable predictions. In cases with imaging data included for the study (e.g., lung scans for COVID-19), CNN models are developed using python libraries like TensorFlow and Keras to process image data. The advantage of image processing algorithms that are part of these libraries is that they are effective in providing better accuracy in detecting infection patterns from data (Ardakani et al. 2020).

Genetic and Rare Diseases: Genomic data is a high-dimensional data that requires advanced tools for processing. Advanced AI techniques like DL helps in providing accurate analysis. Python libraries TensorFlow and Keras are optimal for working with deep neural network algorithms for the processing of these large genomic datasets. Ensemble methods and embedded feature selection techniques are also applied for reducing dimensionality and enhancing model interpretability (Zou and Hastie 2005).

6.4 Comparing Performance and Applicability in Clinical Settings

Clinical application and acceptance of AI tools are based on performance and its evaluation based on various evaluation matrices. Effective application use of the AI tools is required for prediction and the interpretation of results. The easy integration of these tools is needed for real-time application in clinical workflows. Interpretability and scalability are the core factors that help with the success of these tools. Deep learning models built by the python libraries TensorFlow and Keras outperform when compared with simpler models in terms of accuracy. The lack of transparency in these models can hinder the adoption of tools in clinical settings, where interpretability is crucial (Topol 2019). The models built using python libraries Scikit-Learn are less complex and provide higher interpretability and quicker prototyping. The models built with Scikit-learn make them suitable for applications that need transparency. Healthcare systems prefer models built on platforms like IBM Watson Health, which combines AI with clinical data integration capabilities. These tools also comply with healthcare regulations supporting quicker deployment in real-world environments. Proprietary tools may lack flexibility and limit the customization allowed when compared to open-source alternatives.

7. Conclusion

Disease prediction systems have been advanced to provide precise personalized treatment strategies. Advanced tools help precision medicine in early diagnosis and preparation of personalized treatment plans. This chapter has discussed various tools and techniques that help with building AI tools for precision patient care. ML, DL, data mining and statistical modeling are various sets of tools that contribute with the development of advanced disease prediction strategies. The multimodal nature of healthcare data accommodates various medical conditions. With the selection of appropriate tools and techniques for the modelling, AI tools help with the development of effective, accurate, and clinically applicable AI models and tools.

Important factors that are to be kept in mind while building disease prediction models include effective feature selection, data preprocessing, and model evaluation. The feature selection process helps with the reduction of complexity from the multimodal data and enhances the model interpretability. The feature selection process helps in focusing the predictions based on the most informative variables. Data preprocessing is essential with managing healthcare data and its challenges. The preprocessing techniques are used for improving data quality by ensuring consistency, and optimizing model performance. Comprehensive model evaluation is done using performance evaluation metrics like AUC-ROC, F1 score, and cross-validation techniques. Model evaluation helps by ensuring the models perform better with various data sets, ensuring that the models also accommodate diverse patient populations, by delivering reliable predictions in clinical applications.

Python libraries like Scikit-Learn, TensorFlow, and Tools like IBM Watson Health and its application reveal that each of the tools has its own advantages. These tools help to process multimodal healthcare data. The tools can handle various models, complexity, interpretability, and integration requirements. Open-source python libraries like Scikit-Learn and TensorFlow provide flexibility and control when working with custom model developments. Specialized healthcare tools like IBM Watson Health help with robust deployment features that are suited for clinical applications and limits customization. These tools help the developers and practitioners in aligning their choice of tools and techniques with specific disease characteristics based on data availability and clinical applications.

There is no single tool or technique that can fit and process all disease predictions. The optimal approach for selecting and identifying these tools depends on the need. This includes disease type, dataset structure, and the clinical priorities. Accuracy and interpretability are crucial factors that are to be considered while designing and developing clinical tools. Future advancements in disease prediction increase the integration of multimodal data and improve transparency of DL models. The expansion of new techniques like federated learning help to accommodate privacy constraints in healthcare data sharing. Careful consideration of strengths and limitations of tools and techniques is required to identify the efficiency of the tools. Advanced healthcare systems are to be developed to create predictive models that support informed clinical decision-making, enhance patient outcomes, and contribute to a personalized treatment approach in healthcare.

References

Abadi, Martín, Paul Barham, Jianmin Chen, Zhifeng Chen, Andy Davis, Jeffrey Dean et al. (2016). {TensorFlow}: a system for {Large-Scale} machine learning. pp. 265–283. In 12th USENIX Symposium on Operating Systems Design and Implementation (OSDI 16).

Alanazi, Rayan. (2022). Identification and prediction of chronic diseases using machine learning approach. Journal of Healthcare Engineering 1: 2826127.

Ardakani, Ali Abbasian, Alireza Rajabzadeh Kanafi, Rajendra Acharya, U., Nazanin Khadem and Afshin Mohammadi. (2020). Application of deep learning technique to manage COVID-19 in routine clinical practice using CT images: Results of 10 convolutional neural networks. Computers in Biology and Medicine 121: 103795.

Azur, Melissa J., Elizabeth A. Stuart, Constantine Frangakis and Philip J. Leaf. (2011). Multiple imputation by chained equations: what is it and how does it work? International Journal of Methods in Psychiatric Research 20(1): 40–49.

Bonato, Vinicius, Veerabhadran Baladandayuthapani, Bradley M. Broom, Erik P. Sulman, Kenneth D. Aldape and Kim-Anh Do. (2011). Bayesian ensemble methods for survival prediction in gene expression data. Bioinformatics 27(3): 359–367.

Bradley, Andrew P. (1997). The use of the area under the ROC curve in the evaluation of machine learning algorithms. Pattern Recognition 30(7): 1145–1159.

Breiman, Leo. (2001). Random forests. Machine Learning 45: 5–32.

Cabello-Solorzano, Kelsy, Isabela Ortigosa de Araujo, Marco Peña, Luís Correia and Antonio J. Tallón-Ballesteros. (2023). The impact of data normalization on the accuracy of machine learning algorithms: a comparative analysis. pp. 344–353. In International Conference on Soft Computing Models in Industrial and Environmental Applications. Cham: Springer Nature Switzerland.

Chandrashekar, Girish and Ferat Sahin. (2014). A survey on feature selection methods. Computers & Electrical Engineering 40(1): 16–28.

Chicco, Davide and Giuseppe Jurman. (2020). The advantages of the Matthews correlation coefficient (MCC) over F1 score and accuracy in binary classification evaluation. BMC Genomics 21: 1–13.

Chouhan, Vikash, Sanjay Kumar Singh, Aditya Khamparia, Deepak Gupta, Prayag Tiwari, Catarina Moreira et al. (2020). A novel transfer learning based approach for pneumonia detection in chest X-ray images. Applied Sciences 10(2): 559.

Christen, Peter, David J. Hand and Nishadi Kirielle. (2023). A review of the F-measure: its history, properties, criticism, and alternatives. ACM Computing Surveys 56(3): 1–24.

Davis, Jesse and Mark Goadrich. (2006). The relationship between Precision-Recall and ROC curves. pp. 233–240. In Proceedings of the 23rd International Conference on Machine Learning.

Esteva, Andre, Alexandre Robicquet, Bharath Ramsundar, Volodymyr Kuleshov, Mark DePristo, Katherine Chou et al. (2019). A guide to deep learning in healthcare. Nature Medicine 25(1): 24–29.

García, Salvador, Julián Luengo and Francisco Herrera. (2015). Data Preprocessing in Data Mining. Vol. 72. Cham, Switzerland: Springer International Publishing.

Goldstein, Benjamin A. and Michael J. Pencina. (2014). Developing implementable risk prediction models with electronic health records data. Wiley statsRef: Statistics Reference Online 1–8.

Goldstein, Benjamin A., Ann Marie Navar, Michael J. Pencina and John P. A. Ioannidis. (2017). Opportunities and challenges in developing risk prediction models with electronic health records data: a systematic review. Journal of the American Medical Informatics Association: JAMIA 24(1): 198.

Guyon, Isabelle and André Elisseeff. (2003). An introduction to variable and feature selection. Journal of Machine Learning Research 3(Mar): 1157–1182.

Hall, Mark A. (1999). Correlation-based Feature Selection for Machine Learning. PhD diss., The University of Waikato.

Hamilton, Alisa J., Alexandra T. Strauss, Diego A. Martinez, Jeremiah S. Hinson, Scott Levin, Gary Lin et al. (2021). Machine learning and artificial intelligence: applications in healthcare epidemiology. Antimicrobial Stewardship & Healthcare Epidemiology 1(1): e28.

Hastie, Trevor, Robert Tibshirani and Jerome Friedman. (2017). The elements of statistical learning: data mining, inference, and prediction. .

Jiang, Fei, Yong Jiang, Hui Zhi, Yi Dong, Hao Li, Sufeng Ma et al. (2017). Artificial intelligence in healthcare: past, present and future. Stroke and Vascular Neurology 2(4).

Jung, Sin-Ho. (2014). Stratified Fisher's exact test and its sample size calculation. Biometrical Journal 56(1): 129–140.

Kaushik, Shruti, Abhinav Choudhury, Pankaj Kumar Sheron, Nataraj Dasgupta, Sayee Natarajan, Larry A. Pickett et al. (2020). AI in healthcare: time-series forecasting using statistical, neural, and ensemble architectures. Frontiers in Big Data 3: 4.

Kohavi, R. (1995). A Study of Cross-validation and Bootstrap for Accuracy Estimation and Model Selection. Morgan Kaufman Publishing.

Kohavi, Ron and George H. John. (1997). Wrappers for feature subset selection. Artificial Intelligence 97(1-2): 273–324.

Kuhn, M. (2013). Applied predictive modeling..

Li, Jundong, Kewei Cheng, Suhang Wang, Fred Morstatter, Robert P. Trevino, Jiliang Tang et al. (2017). Feature selection: A data perspective. ACM Computing Surveys (CSUR) 50(6): 1–45.

Litjens, Geert, Thijs Kooi, Babak Ehteshami Bejnordi, Arnaud Arindra Adiyoso Setio, Francesco Ciompi, Mohsen Ghafoorian et al. (2017). A survey on deep learning in medical image analysis. Medical Image Analysis 42: 60–88.

Little, Roderick J. A. and Donald B. Rubin. (2019). Statistical Analysis with Missing Data. Vol. 793. John Wiley & Sons.

Liu, Fei Tony, Kai Ming Ting and Zhi-Hua Zhou. (2008). Isolation forest. pp. 413–422. In 2008 Eighth IEEE International Conference on Data Mining. IEEE.

Liu, Huan and Hiroshi Motoda. (2012). Feature Selection for Knowledge Discovery and Data Mining. Vol. 454. Springer Science & Business Media.

Nayak, Veerabhadra Sanekal, Mohammed Saleem Khan, Bharat Kumar Shukla and Pranjal R. Chaturvedi. (2016). Artificial intelligence in clinical research. Int J Clin Trials 3(4): 187–193.

Obermeyer, Ziad and Ezekiel J. Emanuel. (2016). Predicting the future—big data, machine learning, and clinical medicine. New England Journal of Medicine 375(13): 1216–1219.

Pedregosa, Fabian, Gaël Varoquaux, Alexandre Gramfort, Vincent Michel, Bertrand Thirion, Olivier Grisel et al. (2011). Scikit-learn: Machine learning in Python. The Journal of Machine Learning Research 12: 2825–2830.

Poongodi, T., Sumathi, D., Suresh, P. and Balamurugan Balusamy. (2021). Deep learning techniques for electronic health record (EHR) analysis. Bio-inspired Neurocomputing 73–103.

Powers, David M. W. (2020). Evaluation: from precision, recall and F-measure to ROC, informedness, markedness and correlation. arXiv preprint arXiv:2010.16061.

Rajpurkar, Pranav, Jeremy Irvin, Robyn L. Ball, Kaylie Zhu, Brandon Yang, Hershel Mehta et al. (2018). Deep learning for chest radiograph diagnosis: A retrospective comparison of the CheXNeXt algorithm to practicing radiologists. PLoS Medicine 15(11): e1002686.

Sajjadnia, Zeinab, Raof Khayami and Mohammad Reza Moosavi. (2020). Preprocessing breast cancer data to improve the data quality, diagnosis procedure, and medical care services. Cancer Informatics 19: 1176935120917955.

Secinaro, Silvana, Davide Calandra, Aurelio Secinaro, Vivek Muthurangu and Paolo Biancone. (2021). The role of artificial intelligence in healthcare: a structured literature review. BMC Medical Informatics and Decision Making 21: 1–23.

Shickel, Benjamin, Patrick James Tighe, Azra Bihorac and Parisa Rashidi. (2017). Deep EHR: a survey of recent advances in deep learning techniques for electronic health record (EHR) analysis. IEEE Journal of Biomedical and Health Informatics 22(5): 1589–1604.

Sokolova, Marina, Nathalie Japkowicz and Stan Szpakowicz. (2006). Beyond accuracy, F-score and ROC: a family of discriminant measures for performance evaluation. pp. 1015–1021. In Australasian Joint Conference on Artificial Intelligence. Berlin, Heidelberg: Springer Berlin Heidelberg.

Spiegelhalter, David J., Keith R. Abrams and Jonathan P. Myles. (2004). Bayesian Approaches to Clinical Trials and Health-Care Evaluation. Vol. 13. John Wiley & Sons.

Tharwat, Alaa. (2021). Classification assessment methods. Applied Computing and Informatics 17(1): 168–192.

Tibshirani, Robert. (1996). Regression shrinkage and selection via the lasso. Journal of the Royal Statistical Society Series B: Statistical Methodology 58(1): 267–288.

Topol, Eric J. (2019). High-performance medicine: the convergence of human and artificial intelligence. Nature Medicine 25(1): 44–56.

Van der Maaten, Laurens and Geoffrey Hinton. (2008). Visualizing data using t-SNE. Journal of Machine Learning Research 9(11).
Wold, Svante, Kim Esbensen and Paul Geladi. (1987). Principal component analysis. Chemometrics and Intelligent Laboratory Systems 2(1-3): 37–52.
Zeger, Scott L., Rafael Irizarry and Roger D. Peng. (2006). On time series analysis of public health and biomedical data. Annu. Rev. Public Health 27(1): 57–79.
Zou, Hui and Trevor Hastie. (2005). Regularization and variable selection via the elastic net. Journal of the Royal Statistical Society Series B: Statistical Methodology 67(2): 301–320.

9

Applications of Smart Systems for Disease Prediction

1. Introduction

Integrating Artificial Intelligence (AI) and smart systems in healthcare have transformed traditional disease diagnosis, treatment, and prevention approaches. Smart systems, encompassing a range of AI-driven technologies, are increasingly critical in predicting diseases and enabling healthcare providers to make proactive, informed decisions. These systems harness vast amounts of health data—from electronic health records (EHRs), medical imaging, wearable sensors, genomics, and patient-reported outcomes—facilitating insights into disease patterns, patient risk factors, and treatment responses (Topol 2019). This paradigm shift towards predictive and preventive healthcare holds particular promise in managing chronic diseases, that often require early detection and consistent monitoring to prevent complications and improve patient outcomes.

Early disease prediction has emerged as a key area of focus within intelligent healthcare systems. Early detection is crucial, especially for conditions like cancer, cardiovascular diseases, and neurological disorders, where prompt intervention can significantly reduce morbidity and mortality rates. Predictive models trained on historical health data identify high-risk patients before symptoms become apparent, supporting early diagnosis and enabling timely, personalized interventions (Hider et al. 2024). For instance, predictive analytics have shown effectiveness in identifying early signs of diabetes-related complications, allowing for tailored interventions that mitigate long-term damage and healthcare costs (Neravetla et al. 2024). These applications underscore the shift from a reactive to a proactive healthcare model, prioritizing prevention and early management.

In precision medicine, AI-based smart systems enable a nuanced approach to treatment, customizing recommendations to each patient's unique genetic,

DOI: 10.1201/9781032722344-9

behavioral, and environmental profile. Predictive analytics tools in this field analyze various data layers to forecast disease progression and therapy response, moving away from generalized treatment models. Such AI applications exemplify the power of data-driven healthcare, which optimizes resources and enhances patient care (Collins and Varmus 2015). For example, predictive algorithms can identify genetic markers linked to cancer susceptibility, guiding targeted screening and personalized treatment plans (Tajbakhsh et al. 2016).

Smart systems for disease prediction leverage a range of AI methodologies, including machine learning (ML), deep learning (DL), and natural language processing (NLP), each offering distinct advantages. ML models process structured data such as lab results, demographic information, and clinical history to derive predictive patterns. In contrast, DL models, particularly convolutional neural networks (CNNs), excel at analyzing unstructured data like medical images (Litjens et al. 2017). Additionally, NLP algorithms interpret unstructured text data from clinical notes, pathology reports, and scientific literature, extracting valuable insights that can enrich predictive models. By integrating these diverse data sources and AI techniques, smart systems offer holistic insights into disease development, identifying high-risk individuals and providing real-time diagnostic support.

This chapter delves into the diverse applications of smart systems within disease prediction, highlighting their role in various domains, including cardiovascular diseases, cancer, diabetes, neurological disorders, and infectious diseases. Focusing on early prediction applications, the chapter explores how these technologies assist clinicians in making accurate, data-backed predictions that enhance patient outcomes. Each section provides examples of practical applications, detailing the techniques used, the data leveraged, and the impact on healthcare delivery. Concluding with an analysis of specific smart systems for high-impact diseases, this chapter aims to elucidate the transformative potential of AI-enabled predictive models in shaping the future of healthcare.

2. Smart Systems in Disease Prediction

The evolution of smart systems has significantly advanced the field of disease prediction, allowing healthcare providers to identify early signs of disease and prevent adverse health outcomes. By leveraging a combination of AI, ML, DL, and data analytics, these systems transform raw health data into actionable insights. Smart systems for disease prediction utilize structured and unstructured data from multiple sources, including electronic health records (EHRs), medical imaging, wearable devices, genomics, and clinical research, enabling a multidimensional approach to health risk assessment and management (Esteva et al. 2019). The key applications of smart systems for various diseases are been summerised in Figure 1 and Table 1.

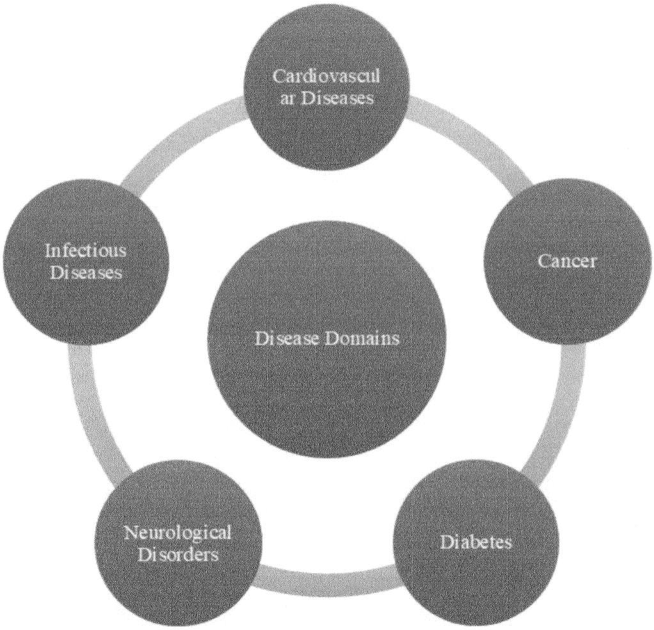

Figure 1. Key applications of smart systems across disease domains.

Table 1. Key applications of smart systems across disease domains.

Disease Domain	Data Sources	AI Techniques Used	Outcomes
Cardiovascular Diseases	EHRs, Wearables, Imaging	Logistic Regression, CNNs, NLP	Early detection, risk prediction, real-time monitoring
Cancer	Medical Imaging, Genomics	CNNs, Survival Analysis Models	Early tumor detection, prognosis, personalized therapy
Diabetes	CGM, EHRs, Lifestyle Data	RNNs, LSTMs	Glucose trend prediction, automated insulin delivery
Neurological Disorders	MRI, PET Scans, Wearable Data	CNNs, Multi-modal Neural Networks	Early detection of Alzheimer's, Parkinson's, cognitive decline risks
Infectious Diseases	Surveillance Data, Social Media	Time-Series Analysis, NLP	Outbreak prediction, real-time surveillance

2.1 Architecture of Smart Disease Prediction Systems

A typical smart disease prediction system follows a layered architecture, combining data ingestion, processing, model training, and prediction modules. Data ingestion involves extracting relevant information from diverse data sources, which is then processed through data-cleaning, normalization, and transformation steps to prepare it for machine learning models. Advanced data-processing techniques, including natural language processing (NLP) and image preprocessing, allow these systems to work with both structured data (e.g., laboratory test results) and unstructured data (e.g., clinical notes and imaging).

Model training is central to the disease prediction process, where algorithms learn from historical data to detect patterns indicative of disease onset or progression. Smart systems for disease prediction often incorporate various ML and DL algorithms, with each serving a distinct purpose: while ML models such as logistic regression or decision trees provide interpretable outputs suitable for clinical settings, DL models like convolutional neural networks (CNNs) and recurrent neural networks (RNNs) process complex data formats like images and time-series data, enhancing prediction accuracy (Gulshan et al. 2016).

2.2 Data Processing and Integration

Data integration is a critical component of disease prediction systems, as healthcare data often resides in disparate formats and silos. Effective disease prediction requires harmonizing data from multiple sources, including EHRs, genomics, imaging data, and sensor data from wearables. For instance, imaging data from MRIs or CT scans must be processed using specialized preprocessing techniques, including normalization, noise reduction, and alignment with anatomical landmarks to ensure consistency (Litjens et al. 2017). Genomic data, on the other hand, requires sequence alignment and variant calling techniques to identify genetic markers associated with disease risk.

In addition, NLP techniques have proven essential for parsing unstructured data, such as physician notes and pathology reports, to extract valuable health information. For example, NLP systems can analyze patient histories to identify potential risk factors or symptoms, contributing to the predictive model's overall accuracy (Hider et al. 2024). This multi-source data integration enables a comprehensive view of patient health, creating predictive models that are more accurate and clinically relevant.

2.3 AI Techniques and Algorithms for Disease Prediction

Smart systems use a diverse array of AI techniques tailored to the specific requirements of each disease type and data format. Key methods include:

Machine Learning (ML): Classical ML models such as Logistic Regression, Decision Trees, and Random Forests provide foundational approaches for disease prediction. These algorithms analyze structured data from EHRs, laboratory results, and demographic information to create interpretable models for clinicians. For instance, logistic regression models are widely used to predict the likelihood of cardiovascular diseases based on patient risk factors (Wouters et al. 2023).

Deep Learning (DL): DL algorithms, particularly CNNs, are highly effective in medical image analysis, such as detecting tumors in radiology images or identifying neurological abnormalities. CNNs automatically learn spatial hierarchies from imaging data, enabling precise disease prediction from medical images. For time-series data, such as glucose monitoring in diabetes, recurrent neural networks

Table 2. AI techniques and their applications in disease prediction.

AI Technique	Application Area	Example Use Case
Logistic Regression	Cardiovascular Risk Assessment	Framingham model for predicting heart attack risk
Convolutional Neural Networks (CNNs)	Cancer Detection	Identifying tumors in mammography images
Recurrent Neural Networks (RNNs)	Diabetes Management	Predicting glucose fluctuations from CGM data
Natural Language Processing (NLP)	Unstructured Data Analysis	Extracting symptoms from clinical notes
Time-Series Forecasting	Infectious Disease Surveillance	Predicting flu outbreak trends

(RNNs) and long short-term memory (LSTM) networks have shown efficacy by capturing temporal patterns in physiological data (Nomura et al. 2021).

Natural Language Processing (NLP): NLP techniques play a critical role in analyzing unstructured text data, such as medical notes, discharge summaries, and research publications. NLP algorithms extract relevant information, such as disease symptoms, risk factors, and diagnosis details, contributing valuable features to predictive models. Advanced NLP models, like Bidirectional Encoder Representations from Transformers (BERT), enhance disease prediction accuracy by interpreting complex linguistic nuances within the clinical text (Kenton et al. 2019).

Some of the AI techniques and their applications for disease prediction have been summarized in Table 2.

2.4 Impact of Smart Systems on Healthcare Practices

The application of smart systems in disease prediction has ushered in significant changes in healthcare, particularly in early intervention, personalized care, and resource optimization. By providing predictive insights, these systems enable healthcare providers to identify high-risk patients before symptoms become apparent, leading to earlier and more effective interventions. For instance, predictive analytics in cardiovascular care allows clinicians to target preventive measures for patients with a high risk of heart disease, thus reducing the likelihood of severe outcomes like myocardial infarction or stroke (Shu et al. 2021).

In oncology, AI-driven prediction systems aid in early cancer detection, allowing for quicker diagnosis and treatment initiation. Studies show that cancer prediction models leveraging AI algorithms have reduced diagnostic errors and improved survival rates, particularly for cancers that are difficult to detect in their early stages (Esteva et al. 2017). In diabetes management, continuous glucose monitoring (CGM) combined with AI models enables real-time prediction of glucose levels, assisting patients in avoiding hypo- or hyperglycemic episodes. Integrating smart prediction systems into daily health management exemplifies the shift toward precision medicine, where treatments and interventions are personalized to each patient's unique health profile (Giorgini et al. 2024).

Moreover, these smart systems have proved valuable in managing healthcare resources. Predictive models help hospitals anticipate patient influx based on seasonal disease trends, allowing them to allocate staff and resources more efficiently. This optimization is particularly relevant in infectious disease management, where predictive analytics can provide early warnings of potential outbreaks, guiding public health interventions (Mishra et al. 2022). By bridging the gap between raw health data and actionable insights, smart systems for disease prediction represent a transformative step toward proactive, data-driven healthcare.

3. Intelligent Systems for Early Disease Prediction

Early disease prediction is one of the most critical applications of AI in healthcare, allowing clinicians to diagnose and manage diseases before they progress to advanced stages. Intelligent systems enable the early identification of health risks by analyzing patterns in patient data and drawing correlations that might not be immediately apparent to human practitioners. These predictive systems are particularly valuable for diseases where early intervention dramatically improves patient outcomes, such as cancer, cardiovascular disease, and diabetes. By leveraging data from various sources, including electronic health records (EHRs), medical imaging, and wearable devices, these systems detect early signs of disease progression, supporting preventive healthcare and personalized treatment strategies (Topol 2019).

3.1 Prognosis and Risk Prediction

The first stage in disease prediction involves assessing a patient's risk of developing specific conditions based on historical health data, lifestyle factors, and genetic predispositions. Prognostic models help clinicians predict a patient's likelihood of developing diseases like cardiovascular disease (CVD) or type 2 diabetes within a specific timeframe. For instance, AI-powered risk prediction models have been developed using data from large-scale epidemiological studies such as the Framingham Heart Study, which provides risk scores based on age, blood pressure, cholesterol levels, and other factors (D'Agostino et al. 2013).

For cardiovascular health, AI-driven risk prediction systems employ ML algorithms that combine EHR data, wearable sensor data, and demographic information. These models predict future cardiovascular events, enabling proactive interventions. Similarly, in oncology, risk prediction models analyze genetic markers and lifestyle factors to estimate the likelihood of developing various types of cancer, allowing individuals at high risk to undergo early screening and surveillance (Tajbakhsh et al. 2016).

3.2 Early Detection Algorithms

In early disease detection, intelligent systems identify markers that indicate disease onset before symptomatic presentation. ML and DL algorithms analyze biomarkers, imaging data, and physiological signals to recognize subtle changes associated with disease progression. For example, convolutional neural networks (CNNs) have been instrumental in detecting early signs of cancers in imaging modalities, such as mammography, CT scans, and MRI (Litjens et al. 2017). These AI-driven systems support radiologists by highlighting potential areas of concern, leading to earlier and more accurate diagnoses.

In diabetes care, continuous glucose monitoring (CGM) devices integrated with AI algorithms provide real-time insights into glucose patterns, predicting potential hyperglycemic or hypoglycemic events before they occur. Recurrent neural networks (RNNs) and long short-term memory (LSTM) networks are especially effective in analyzing time-series data from CGM, enabling diabetes patients to take preventive measures based on predicted glucose levels (Giorgini et al. 2024).

Another notable application of early detection algorithms is in neurological disease prediction. Alzheimer's disease and other neurodegenerative conditions benefit from AI systems that analyze MRI and PET scans to identify early brain structural changes linked to these conditions. Predictive models trained on neuroimaging data detect small variations in brain morphology, enabling diagnosis well before symptoms manifest, which is critical for early intervention (Gill et al. 2020).

3.3 Precision Medicine and Predictive Analytics

Precision medicine aims to customize healthcare recommendations to individual patients based on their unique genetic, environmental, and lifestyle factors. AI-enhanced predictive analytics plays a significant role in precision medicine, guiding personalized care plans that consider an individual's susceptibility to specific diseases and likely responses to treatment. By analyzing complex interactions within biological and environmental data, predictive models can identify patient-specific risk factors and inform personalized treatment strategies.

For example, precision oncology uses predictive analytics to select targeted therapies based on a tumor's genetic profile, improving treatment outcomes and minimizing side effects. ML algorithms analyze genetic mutations, protein expressions, and tumor characteristics to determine the most effective treatment regimens for individual patients. Similarly, in diabetes care, precision medicine approaches use AI models that integrate genetic, metabolic, and lifestyle data to predict the efficacy of specific treatments, such as oral hypoglycemics or insulin therapy, allowing clinicians to tailor interventions accordingly (Vettoretti et al. 2020).

Predictive analytics support precision medicine in cardiovascular medicine by identifying high-risk patients for adverse cardiovascular events, allowing for personalized preventive measures. For instance, AI models that assess patient-specific factors, such as lipid levels and physical activity data, enable physicians to make data-driven recommendations for diet, exercise, and medication to reduce the likelihood of cardiovascular incidents (Shu et al. 2021).

3.4 Applications Across Disease Domains

Cardiovascular Diseases: AI-powered early detection models analyze EHR data, sensor data from wearable devices, and imaging data to detect early indicators of cardiovascular disease, including hypertension and atherosclerosis. By assessing cardiovascular risk factors early, these systems enable proactive intervention, reducing the risk of severe events like heart attack and stroke (Wouters et al. 2023).

Cancer Detection: In oncology, intelligent systems use deep learning to detect early cancer indicators in radiology images, histopathological samples, and genetic data. CNNs, for instance, can identify microscopic features of malignant cells, supporting the early diagnosis of cancers such as breast, lung, and skin cancer. AI models analyze genetic mutations linked to increased cancer susceptibility for high-risk patients, guiding early screening protocols and preventive measures (Esteva et al. 2017).

Diabetes Management: For diabetes, AI models analyze continuous glucose monitoring (CGM) data, lifestyle factors, and historical glucose levels to predict glucose fluctuations and detect prediabetic conditions. These predictions allow timely intervention and support patients in maintaining blood sugar control, thus preventing complications like diabetic neuropathy and retinopathy (Vettoretti et al. 2020).

Neurological Disorders: In neurodegenerative diseases, predictive models analyze MRI and PET scans for early detection of structural brain changes associated with conditions such as Alzheimer's and Parkinson's disease. These systems provide insights into disease progression, enabling early intervention strategies and informing clinical decisions on preventive care and lifestyle adjustments (Hojjati et al. 2022).

Infectious Disease Surveillance: AI-driven disease surveillance systems analyze real-time data from healthcare facilities, social media, and environmental sources to detect patterns indicative of infectious disease outbreaks. Predictive models for infectious diseases, such as influenza, combine ML techniques with epidemiological data to provide early warnings, enabling healthcare agencies to implement timely preventive measures and control the spread (Chae et al. 2018).

Intelligent systems for early disease prediction reshape the healthcare landscape, enabling clinicians to intervene before diseases progress to critical stages. Through advancements in ML, DL, and data integration, these systems can analyze complex datasets and identify predictive markers that inform patient care decisions. As AI continues to evolve, the potential for early prediction applications will expand, driving a new era of proactive, precision healthcare that improves patient outcomes and optimizes healthcare resource utilization. By supporting early intervention and personalized care, intelligent disease prediction systems offer a transformative approach to managing and preventing diseases in various medical fields.

4. Applications of Smart Systems by Disease Domain

Smart systems in healthcare leverage AI and ML techniques to provide data-driven insights across diverse disease domains. These systems improve diagnostic accuracy, guide treatment, and support early intervention by applying predictive models tailored to specific conditions. This section explores the applications of smart systems in cardiovascular disease, cancer, diabetes, neurological disorders, and infectious diseases, each highlighting unique methodologies and impacts.

4.1 Cardiovascular Disease Prediction

Cardiovascular diseases (CVD), including myocardial infarction, stroke, and heart failure, are major causes of mortality worldwide. Early prediction and prevention of these diseases are critical for reducing morbidity and mortality rates. Smart systems for cardiovascular disease prediction analyze clinical data from electronic health records (EHRs), wearable devices, and imaging studies to assess patient risk and guide preventive measures. For example, AI-powered models based on supervised learning algorithms, such as Logistic Regression and Random Forests, evaluate cardiovascular risk factors, including cholesterol levels, blood pressure, and lifestyle factors (Wouters et al. 2023).

One prominent application of smart systems in cardiovascular care is in electrocardiogram (ECG) analysis. Deep learning models, particularly convolutional neural networks (CNNs), have been applied to ECG data to identify patterns indicative of arrhythmias and predict the likelihood of adverse cardiovascular events. These models provide real-time monitoring and alerting capabilities, aiding in the early identification of high-risk patients. Additionally, wearable devices such as smartwatches integrate AI algorithms that monitor heart rate and rhythm, offering continuous risk assessment for patients with atrial fibrillation or heart failure (Shu et al. 2021).

4.2 Cancer Detection and Prognosis

Cancer remains a complex and challenging disease domain, where early detection is essential for improving survival rates. Smart systems in oncology employ ML and DL algorithms to detect cancers at an early stage and predict disease progression. Medical imaging, such as mammography, CT scans, and MRI, is a primary data source for cancer prediction, with CNNs demonstrating exceptional performance in identifying malignant patterns within these images (Litjens et al. 2017).

For breast cancer detection, for example, CNN models trained on large imaging datasets can accurately identify tumors, often surpassing human radiologists in diagnostic sensitivity. Beyond detection, AI models assist in predicting cancer prognosis by analyzing tumor histology, molecular markers, and genetic data. In lung cancer, for instance, predictive algorithms assess the likelihood of recurrence based on genomic signatures, supporting personalized treatment plans (Tajbakhsh et al. 2016). Additionally, precision oncology applications use genetic data to identify actionable mutations, enabling targeted therapies for patients with specific genetic profiles (Esteva et al. 2017).

4.3 Diabetes Management and Complication Prediction

Diabetes management involves continuous monitoring of blood glucose levels to prevent complications such as neuropathy, retinopathy, and cardiovascular disease. Smart systems in diabetes care use predictive analytics and real-time monitoring to assist patients in maintaining stable glucose levels. Continuous glucose monitoring (CGM) devices integrated with AI algorithms analyze blood glucose data and provide alerts for impending hypoglycemic or hyperglycemic episodes. Recurrent neural networks (RNNs) and long short-term memory (LSTM) models are commonly used to analyze CGM data, allowing the prediction of glucose trends over time (Giorgini et al. 2024).

Smart insulin delivery systems, such as closed-loop insulin pumps, use predictive models to adjust insulin administration in real-time, based on glucose level forecasts, effectively acting as an artificial pancreas. In addition, smart systems predict diabetes-related complications by analyzing patient data from EHRs and wearable devices, identifying individuals at high risk for cardiovascular or renal complications associated with diabetes (Vettoretti et al. 2020). These AI-powered systems enable personalized diabetes management, helping patients and clinicians take preventive actions based on dynamic glucose predictions.

4.4 Neurological Disorder Prediction

Neurodegenerative diseases, such as Alzheimer's disease and Parkinson's disease, present significant challenges due to their progressive nature and limited treatment options. Smart systems in neurology support early detection by analyzing neuroimaging data from MRI, CT, and PET scans, allowing clinicians to identify structural and functional changes in the brain associated with disease onset. Predictive models trained on imaging data help differentiate between neurodegenerative diseases and typical aging processes, enhancing diagnostic accuracy in the early stages (Gill et al. 2020).

For Alzheimer's disease, deep learning algorithms analyze brain MRI scans to detect subtle changes in brain volume and cortical thickness, which are early indicators of the disease. Additionally, AI models predict cognitive decline by analyzing genetic, biomarker, and lifestyle data, allowing for personalized intervention strategies. In Parkinson's disease, smart systems analyze gait patterns and other motor symptoms captured by wearable devices, detecting early motor dysfunction indicative of the disease (Hojjati et al. 2022). These AI-driven predictive models support early interventions, such as cognitive therapies or pharmacological treatments, that may slow disease progression.

4.5 Infectious Disease Surveillance and Prediction

Infectious disease outbreaks, including seasonal influenza and emerging epidemics like COVID-19, require rapid detection and containment to prevent widespread transmission. AI-based surveillance systems analyze data from healthcare facilities, public health records, social media, and environmental factors to detect early signs of infectious disease outbreaks. ML models use time-series analysis and natural language processing (NLP) to forecast disease trends and provide early warnings, allowing for timely public health interventions (Chae et al. 2018).

For example, AI models used in flu surveillance analyze historical flu trends, temperature variations, and population mobility data to predict outbreak patterns, supporting vaccination and prevention strategies. In the case of COVID-19, predictive models used data from mobility patterns, testing records, and social media mentions to estimate infection rates, guiding policymakers in implementing targeted restrictions. These smart systems play a vital role in global health security, enabling efficient responses to outbreaks and improving healthcare resource allocation during public health emergencies (Keshavamurthy et al. 2022).

Smart systems tailored to specific disease domains significantly advance early detection, risk assessment, and personalized healthcare. By utilizing large-scale data and sophisticated AI models, these systems enable healthcare providers to proactively manage various conditions, from chronic diseases like cardiovascular disease and diabetes to acute threats like infectious disease outbreaks. As these intelligent systems evolve, their impact on disease prediction and prevention will deepen, supporting the healthcare industry's shift toward a proactive, data-driven care model.

5. Working Principles of Key Disease Prediction Systems

The working principles of disease prediction systems are based on integrating vast amounts of data, applying predictive algorithms, and leveraging specific features relevant to each disease domain. These systems typically involve preprocessing steps, ML or DL algorithms for pattern recognition, and output interpretation to provide actionable insights. This section examines the underlying principles of key prediction systems across cardiovascular diseases, cancer, diabetes, neurological disorders, and infectious diseases.

5.1 Cardiovascular Disease Prediction Systems

Cardiovascular disease prediction systems primarily use structured data from electronic health records (EHRs) and wearable devices combined with demographic and clinical factors. The working principle involves:

Data Preprocessing: The initial step involves cleansing and normalizing data from multiple sources, such as blood pressure, cholesterol levels, ECG data, and lifestyle factors. This preprocessing step ensures consistency and removes noise that may affect model performance.

Predictive Modeling: Supervised ML algorithms, such as Logistic Regression, Decision Trees, and Random Forests, are used to analyze the likelihood of cardiovascular events. Logistic regression models assess linear relationships between risk factors and outcomes, while ensemble methods like Random Forests combine multiple models to improve prediction accuracy (Wouters et al. 2023).

Real-Time Monitoring: Wearable devices continuously capture ECG and heart rate variability data for real-time prediction. Deep learning models, particularly convolutional neural networks (CNNs), are applied to ECG signals to detect arrhythmias and other cardiac anomalies that may indicate high-risk events. By incorporating both historical and real-time data, cardiovascular prediction systems offer continuous monitoring and timely alerts for preventive intervention (Shu et al. 2021).

5.2 Cancer Prognosis and Detection Systems

Cancer prediction systems leverage medical imaging, histopathological samples, and genomic data to identify early cancer markers and predict disease progression. These systems work as follows:

Data Integration and Preprocessing: Cancer data sourced from imaging modalities (MRI, CT scans), pathology reports, and genetic profiling undergoes specialized preprocessing. Imaging data is normalized, noise-filtered, and aligned to enhance the clarity of tumor features. In genomic data, mutation alignment and variant calling are essential to highlight actionable mutations linked to cancer susceptibility (Litjens et al. 2017).

Deep Learning Models: CNNs are widely used in cancer detection, especially for image-based analysis. For example, CNNs trained in radiology images of lung tissue or breast tissue can identify malignant features indicative of cancer. These models extract hierarchical patterns from images, providing detailed insights into tumor structure and aiding radiologists in early diagnosis.

Survival and Recurrence Prediction: Cancer prognosis models, particularly survival analysis models like Cox proportional hazards regression, predict patient outcomes based on tumor characteristics and treatment history. For recurrence prediction, ML models use patient demographics, treatment types, and tumor biology data to estimate the probability of cancer returning, supporting personalized follow-up care (Tajbakhsh et al. 2016).

5.3 Diabetes Management and Complication Prediction Systems

Diabetes prediction systems combine continuous glucose monitoring (CGM) data, wearable devices, and patient records to provide real-time management and complication prediction. Key working principles include:

Time-Series Analysis: CGM devices continuously capture glucose levels, creating a rich time-series data source. Recurrent neural networks (RNNs) and long short-term memory (LSTM) models are effective for predicting future glucose levels, as they capture temporal patterns and trends within the glucose data (Giorgini et al. 2024).

Automated Insulin Delivery: In advanced diabetes systems, predictive algorithms are integrated with closed-loop insulin pumps to automate insulin delivery based on predicted glucose trends. These systems, often called 'artificial pancreas' devices, rely on RNNs and reinforcement learning algorithms to dynamically adjust insulin dosages, improving glycemic control (Vettoretti et al. 2020).

Complication Risk Prediction: ML models analyze patient data, including glucose levels, blood pressure, and lipid profiles, to assess the risk of diabetes-related complications such as retinopathy, neuropathy, and cardiovascular disease. These predictions support clinicians in implementing preventive measures and monitoring high-risk patients more closely.

5.4 Neurological Disorder Early Detection Systems

For neurological disorders like Alzheimer's disease and Parkinson's disease, early detection is essential for managing disease progression. Prediction systems in this domain often rely on neuroimaging data, genetic factors, and biomarkers, working as follows:

Neuroimaging Analysis: MRI and PET scans provide detailed brain structure and function images. Predictive models use deep learning techniques, particularly CNNs, to detect early structural changes such as cortical thinning or hippocampal atrophy associated with Alzheimer's disease. By analyzing neuroimaging data, these models can identify brain alterations linked to disease onset, even before clinical symptoms emerge (Gill et al. 2020).

Genomic and Biomarker Integration: Genomic markers and cerebrospinal fluid (CSF) biomarkers are increasingly used in predicting neurodegenerative diseases. Prediction models integrate biomarker levels with imaging data to improve diagnostic accuracy, often employing multi-modal neural networks that handle both image and numeric data, facilitating a holistic view of patient risk (Hojjati et al. 2022).

Wearable Data for Parkinson's Detection: In Parkinson's disease, smart systems use data from wearable devices to monitor gait, tremors, and other motor symptoms. Algorithms analyze these patterns over time, detecting early signs of motor dysfunction that may indicate the onset of Parkinson's disease. These systems apply signal processing and time-series ML algorithms for real-time monitoring to identify motor fluctuations indicative of disease progression.

5.5 Infectious Disease Outbreak Prediction Systems

Infectious disease prediction systems are vital for managing public health and preventing epidemics. These systems gather data from healthcare facilities, social media, climate models, and transportation networks to provide real-time outbreak surveillance, functioning as follows:

Epidemiological Data Analysis: Epidemiological data from healthcare systems and public health databases provide case counts, demographic information, and geographic spread. ML models, such as time-series forecasting methods, analyze these trends to identify potential outbreaks, informing public health strategies (Chae et al. 2018).

Natural Language Processing (NLP): NLP techniques analyze social media posts, news articles, and other online content to detect early signs of disease outbreaks. These models extract keywords, sentiment, and geolocation data to identify regions where disease prevalence may increase, supplementing traditional surveillance data (Keshavamurthy et al. 2022).

Environmental and Mobility Data Integration: Integrating environmental data (e.g., temperature, humidity) and mobility data (e.g., airline travel, local movement patterns) enables models to predict disease transmission patterns. For example, ML models used in flu and COVID-19 predictions leverage environmental variables to anticipate seasonal peaks and suggest preventive actions. These insights support targeted vaccination campaigns and resource allocation, enabling efficient responses during outbreak scenarios.

Each disease prediction system discussed in this section demonstrates a tailored approach based on the unique characteristics of the disease domain. Through data preprocessing, integration of multi-source data, and application of advanced ML and DL models, these systems provide predictive insights that enhance early diagnosis, guide treatment, and support preventive healthcare strategies. As predictive algorithms and data sources continue to advance, the principles outlined here will evolve, enabling more precise, timely, and accessible disease prediction capabilities across healthcare settings.

6. Conclusion

Integrating smart systems into disease prediction represents a significant leap forward in healthcare, empowering clinicians with tools for early detection, risk assessment, and personalized treatment. By leveraging advanced artificial intelligence (AI) techniques, such as machine learning (ML) and deep learning (DL), these systems can analyze diverse data sources—ranging from electronic health records (EHRs) and medical imaging to genomic information and real-time data from wearable devices. This multi-dimensional approach enables the identification of disease patterns and predictive markers, often before symptoms emerge, which is particularly impactful in domains like oncology, cardiovascular care, diabetes management, and neurology.

As demonstrated across various disease domains, smart systems in healthcare do more than automate tasks; they offer data-driven insights that enhance diagnostic accuracy, guide treatment decisions, and support preventive healthcare measures. In cardiovascular disease, for instance, real-time monitoring through wearable technology and predictive algorithms enables clinicians to identify high-risk patients, potentially averting severe cardiac events. In oncology, AI-driven imaging analysis and genomic profiling allow for early cancer detection and personalized therapies, improving survival rates and reducing treatment-related side effects. In chronic disease management, such as diabetes, smart systems predict glycemic trends and automate insulin delivery, offering more consistent glucose control and reducing long-term complications.

These advancements mark a shift from reactive to proactive healthcare, where disease prediction and prevention are prioritized. By enabling personalized medicine approaches, smart systems support healthcare tailored to each patient's unique characteristics, from genetic predispositions to lifestyle factors. This transition improves patient outcomes and optimizes healthcare resources by focusing on early intervention and disease prevention rather than costly late-stage treatments.

Looking to the future, smart systems for disease prediction are poised to become even more sophisticated as new data sources and AI techniques emerge. Developments in multi-modal AI, federated learning, and explainable AI promise to further enhance predictive models' transparency, privacy, and accuracy. However, implementing these systems in clinical settings raises challenges, such as ensuring data privacy, addressing algorithmic biases, and integrating these technologies into healthcare workflows.

Ultimately, as research and innovation continue, the potential for smart systems in disease prediction will grow, contributing to an increasingly patient-centered, predictive, and preventive healthcare landscape. By bridging the gap between data and actionable insights, these systems promise to transform healthcare into a more proactive, precise, and personalized field—aligning technological advances with the goal of improving lives.

References

Chae, Sangwon, Sungjun Kwon and Donghyun Lee. (2018). Predicting infectious disease using deep learning and big data. International Journal of Environmental Research and Public Health 15(8): 1596.

Collins, Francis S. and Harold Varmus. (2015). A new initiative on precision medicine. New England Journal of Medicine 372(9): 793–795.

D'Agostino Sr, Ralph B., Michael J. Pencina, Joseph M. Massaro and Sean Coady. (2013). Cardiovascular disease risk assessment: insights from Framingham. Global Heart 8(1): 11–23.

Esteva, Andre, Alexandre Robicquet, Bharath Ramsundar, Volodymyr Kuleshov, Mark DePristo, Katherine Chou et al. (2019). A guide to deep learning in healthcare. Nature Medicine 25(1): 24–29.

Esteva, Andre, Brett Kuprel, Roberto A. Novoa, Justin Ko, Susan M. Swetter, Helen M. Blau et al. (2017). Dermatologist-level classification of skin cancer with deep neural networks. Nature 542(7639): 115–118.

Gill, Sascha, Pauline Mouches, Sophie Hu, Deepthi Rajashekar, Frank P. MacMaster, Eric E. Smith et al. (2020). Using machine learning to predict dementia from neuropsychiatric symptom and neuroimaging data. Journal of Alzheimer's Disease 75(1): 277–288.

Giorgini, F., Di Dalmazi, G. and Diciotti, S. (2024). Artificial intelligence in endocrinology: a comprehensive review. Journal of Endocrinological Investigation 47(5): 1067–1082.

Gulshan, Varun, Lily Peng, Marc Coram, Martin C. Stumpe, Derek Wu, Arunachalam Narayanaswamy et al. (2016). Development and validation of a deep learning algorithm for detection of diabetic retinopathy in retinal fundus photographs. Jama 316(22): 2402–2410.

Hider, Mohammad Abir, Md Nasiruddin and Abdullah Al Mukaddim. (2024). Early disease detection through advanced machine learning techniques: a comprehensive analysis and implementation in healthcare systems. Revista de Inteligencia Artificial en Medicina 15(1): 1010–1042.

Hojjati, Seyed Hani, Abbas Babajani-Feremi and Alzheimer's Disease Neuroimaging Initiative. (2022). Prediction and modeling of neuropsychological scores in Alzheimer's disease using multimodal neuroimaging data and artificial neural networks. Frontiers in Computational Neuroscience 15: 769982.

Kenton, Jacob Devlin Ming-Wei Chang and Lee Kristina Toutanova. (2019). Bert: Pre-training of deep bidirectional transformers for language understanding. In Proceedings of naacL-HLT vol. 1, p. 2.

Keshavamurthy, Ravikiran, Samuel Dixon, Karl T. Pazdernik and Lauren E. Charles. (2022). Predicting infectious disease for biopreparedness and response: A systematic review of machine learning and deep learning approaches. One Health 15: 100439.

Litjens, Geert, Thijs Kooi, Babak Ehteshami Bejnordi, Arnaud Arindra Adiyoso Setio, Francesco Ciompi, Mohsen Ghafoorian et al. (2017). A survey on deep learning in medical image analysis. Medical Image Analysis 42: 60–88.

Mishra, Smriti, Ranjan Kumar, Sanjay Kumar Tiwari and Priya Ranjan. (2022). Machine learning approaches in the diagnosis of infectious diseases: a review. Bulletin of Electrical Engineering and Informatics 11(6): 3509–3520.

Neravetla, Anuteja Reddy, Varun Kumar Nomula, Abdul Sajid Mohammed and S. Dhanasekaran. (2024). Implementing AI-driven diagnostic decision support systems for smart healthcare. pp. 1–6. In 2024 15th International Conference on Computing Communication and Networking Technologies (ICCCNT), IEEE.

Nomura, Akihiro, Masahiro Noguchi, Mitsuhiro Kometani, Kenji Furukawa and Takashi Yoneda. (2021). Artificial intelligence in current diabetes management and prediction. Current Diabetes Reports 21(12): 61.

Shu, Songren, Jie Ren and Jiangping Song. (2021). Clinical application of machine learning-based artificial intelligence in the diagnosis, prediction, and classification of cardiovascular diseases. Circulation Journal 85(9): 1416–1425.

Tajbakhsh, Nima, Jae Y. Shin, Suryakanth R. Gurudu, Todd Hurst, R., Christopher B. Kendall, Michael B. Gotway et al. (2016). Convolutional neural networks for medical image analysis: Full training or fine tuning? IEEE Transactions on Medical Imaging 35(5): 1299–1312.

Topol, Eric. (2019). Deep Medicine: How Artificial Intelligence Can Make Healthcare Human Again. Hachette UK.

Vettoretti, Martina, Giacomo Cappon, Andrea Facchinetti and Giovanni Sparacino. (2020). Advanced diabetes management using artificial intelligence and continuous glucose monitoring sensors. Sensors 20(14): 3870.

Wouters, Philippe C., Rutger R. van de Leur, Melle B. Vessies, Antonius M. W. van Stipdonk, Mohammed A. Ghossein, Rutger J. Hassink et al. (2023). Electrocardiogram-based deep learning improves outcome prediction following cardiac resynchronization therapy. European Heart Journal 44(8): 680–692.

10

Challenges and Future Directions with AI in Healthcare

1. Introduction

Artificial Intelligence (AI) has transformed healthcare, bringing innovations that enhance diagnostic accuracy, optimize treatment plans, and improve patient outcomes. In recent years, AI technologies have significantly advanced, leading to applications across numerous medical disciplines, including predictive analytics, imaging diagnostics, personalized medicine, and real-time patient monitoring. By analyzing vast amounts of medical data, AI systems have the potential to streamline processes, enabling early disease detection and individualized treatment strategies that could drastically improve patient care (Topol 2019).

The healthcare sector generates massive and complex datasets that are challenging for traditional analytical approaches to process. AI's ability to handle these large, multidimensional datasets has unlocked new pathways in understanding disease progression, drug response, and patient outcomes. Deep Learning (DL), Machine Learning (ML), and natural language processing (NLP) have been pivotal in extracting valuable insights from unstructured data, such as medical imaging, clinical notes, and genomics. Despite its potential, integrating AI into clinical workflows is challenging. Issues surrounding data privacy, the interpretability of AI models, and ethical concerns regarding AI-driven decision-making have hindered seamless adoption (Shickel et al. 2017).

Recent developments in AI-enabled healthcare reveal the rapid evolution of technologies aimed at predictive modeling and clinical decision support systems. However, implementing AI in healthcare also introduces complexities in data acquisition, quality assurance, and integration, which are crucial for building reliable, scalable solutions. Due to diverse sources, healthcare data often contain missing values, errors, and have varied formats, making it challenging to construct robust models without substantial preprocessing (Miotto et al. 2018, Chen et al. 2020). These

DOI: 10.1201/9781032722344-10

challenges, alongside technical integration with electronic health records (EHRs) and regulatory compliance, present significant obstacles to adopting AI in healthcare.

The ethical implications of using AI in healthcare, such as potential algorithmic bias and lack of interpretability, pose considerable challenges. Patient data privacy and the security of sensitive health information are ongoing concerns as AI systems depend on extensive datasets that could be at risk of breaches or misuse. Ethical frameworks are essential to guide the responsible use of AI in healthcare and ensure equitable, unbiased outcomes. These frameworks are also necessary to address algorithmic biases that may inadvertently disadvantage specific patient populations, thus impeding healthcare equity (Obermeyer et al. 2019).

The future of AI in healthcare hinges on developing frameworks that address these technical, ethical, and regulatory challenges. Achieving this requires collaboration among healthcare providers, data scientists, policymakers, and regulatory bodies. There is a critical need to create standardized protocols and robust guidelines to ensure AI's safe and ethical deployment in clinical environments. This chapter examines the obstacles and considerations, researchers, developers, and clinicians face in adopting AI technologies in healthcare. It further explores the future scope of AI, identifying areas for improvement that could lead to more reliable, interpretable, and ethically sound AI systems capable of transforming disease prediction and patient care.

2. Adoption Challenges in AI for Healthcare

The adoption of AI in healthcare presents substantial challenges that impact its integration, functionality, and overall success in clinical settings. While AI promises to transform healthcare through enhanced diagnosis, personalized treatment plans, and predictive analytics, several technical, logistical, and ethical barriers hinder its widespread use. Key challenges in data integration, model robustness, and workforce readiness highlight the complexity of implementing AI solutions in diverse healthcare environments. This section delves into healthcare providers, data scientists, and policymakers' primary obstacles when adopting AI in healthcare. Table 1 and Figure 1 show the key challenges in AI adoption in healthcare applications.

Figure 1. Key challenges in AI adoption in healthcare.

Table 1. Key challenges in AI adoption in healthcare.

Challenge	Description	Example
Technical Integration	Difficulties in aligning AI with existing healthcare IT systems (e.g., EHRs).	Lack of interoperability between hospital systems.
Data Quality	Issues with data fragmentation, missing values, and inconsistent formats.	Variability in imaging data from different institutions.
Ethical Concerns	Challenges related to algorithmic bias, interpretability, and data privacy.	AI models underperforming for underrepresented demographics.
Workforce Readiness	Insufficient training for healthcare professionals to effectively use AI tools.	Lack of clinician confidence in AI-driven diagnostic tools.
Regulatory Compliance	Complexities in adhering to regional and international regulations for AI in healthcare.	Navigating GDPR or HIPAA for patient data use.

2.1 Technical Complexity and Integration

Integrating AI with existing healthcare IT systems, such as electronic health records (EHRs), is complex and resource-intensive. EHRs often lack standardization across institutions, making it challenging to ensure interoperability and effective data exchange among AI solutions (Birkhead et al. 2015). The technical demands of building, training, and maintaining AI models within the constraints of healthcare infrastructure require extensive computational power, storage capabilities, and reliable access to quality data. Moreover, AI systems must adapt to the diversity of medical records and real-time data streams from varied sources, such as wearable devices, imaging technologies, and lab results, to provide accurate and timely insights (Shortliffe and Sepúlveda 2018).

Another significant challenge is that many AI tools currently function independently rather than as part of an integrated clinical decision-making process. Achieving smooth AI integration requires rethinking hospital workflows to ensure that AI systems align with clinicians' needs without interrupting patient care. This requires coordination between healthcare providers, IT professionals, and AI developers to create systems that seamlessly fit into existing processes (Esteva et al. 2019).

2.2 Data Acquisition and Quality Assurance

The performance of AI models relies heavily on access to high-quality data, which is often difficult to obtain in healthcare due to data fragmentation, missing values, and inconsistent formats. Data collected from different sources and institutions may vary significantly in quality, making it challenging to construct models that generalize effectively across various healthcare settings (Kohane 2015). Missing data, incomplete records, and inaccurate labeling of datasets further complicate model training and can lead to unreliable predictions, ultimately undermining AI's reliability in clinical applications.

Quality assurance processes are essential to ensure that datasets are clean, representative, and reliable. This requires time-intensive data preprocessing, including cleaning, normalization, and validation to remove artifacts and errors. Data quality is especially crucial in healthcare, where inaccurate predictions can have severe consequences. Moreover, since healthcare data often contain complex relationships, such as patient histories and multiple diagnoses, building models that can accurately interpret these intricacies is challenging but necessary for reliable AI systems (Johnson et al. 2016).

2.3 Model Robustness and Reliability

Building AI models that are robust, reliable, and generalizable across diverse patient populations is one of the most pressing challenges in healthcare AI. AI models trained on specific datasets may often not perform accurately when deployed in new settings or on different patient groups. For instance, a model trained on data from one hospital might struggle to produce accurate predictions when applied to another hospital's patient population due to differences in demographics, protocols, and health conditions (Obermeyer et al. 2019). This lack of model robustness can lead to unintended biases in care and limit the applicability of AI systems to varied clinical environments.

Another reliability issue arises from the potential for overfitting, where a model performs exceptionally well on training data but poorly on new, unseen data. Overfitting is particularly concerning in healthcare, where complex patient conditions and unpredictable interactions with treatment regimens demand models that can generalize well. To address these issues, researchers must prioritize robust model validation across multiple datasets and diverse patient populations (Varoquaux 2018).

2.4 Workforce Readiness and Training

A less-discussed but critical barrier to AI adoption in healthcare is workforce readiness. Clinicians, nurses, and healthcare administrators need training to understand and effectively use AI tools in their daily practice. While many AI systems are designed to operate autonomously, clinician oversight is necessary to ensure the appropriate use and interpretation of AI-driven recommendations (Yu et al. 2018). Without adequate training, healthcare professionals may struggle to understand the outputs of AI systems, limiting the effectiveness of these technologies in clinical settings.

Healthcare organizations must prioritize workforce training, helping staff develop technical skills and AI literacy to interact meaningfully with AI tools. Additionally, fostering a culture that values interdisciplinary collaboration between clinicians, data scientists, and IT professionals is essential for maximizing the benefits of AI. Addressing workforce readiness requires training programs that focus on the technical aspects of AI and emphasize the ethical, regulatory, and interpretative challenges that arise in clinical applications.

2.5 Regulatory Compliance and Legal Implications

Compliance with regulatory requirements is a significant hurdle in implementing AI in healthcare. While healthcare regulations, such as the Health Insurance Portability and Accountability Act (HIPAA) in the United States, are designed to protect patient data privacy, they pose challenges for data sharing and access, essential for building high-quality AI models. AI developers and healthcare providers must navigate complex legal frameworks, which often vary by region, to ensure that AI applications comply with relevant healthcare laws (Jiang et al. 2017). This includes ensuring transparency, accountability, and ethical practices throughout the AI lifecycle, from data acquisition to model deployment and monitoring.

The adoption of AI in healthcare faces significant technical, ethical, and regulatory challenges. Overcoming these obstacles requires technological solutions and systemic changes in healthcare practice, regulation, and workforce development. Addressing these challenges is crucial for the safe and effective integration of AI, ultimately leading to improved patient outcomes and healthcare efficiencies.

3. Ethical Considerations

As AI becomes more prevalent in healthcare, ethical considerations grow increasingly important. Ethical challenges stem from the sensitive nature of healthcare data, the high stakes involved in clinical decision-making, and the need for AI systems to operate fairly and transparently. Key ethical concerns include data privacy, algorithmic bias, and the interpretability of AI models. Addressing these issues is critical for fostering trust among healthcare providers and patients and ensuring that AI systems enhance, rather than compromise, healthcare quality and equity.

3.1 Data Privacy and Security

Healthcare data is among the most sensitive forms of personal information, containing details about a patient's health history, treatments, genetic data, and personal identifiers. The ethical imperative to protect patient privacy is paramount, as health data breaches or misuse could lead to severe consequences, including identity theft, discrimination, and psychological harm (Rieke et al. 2020). With AI models relying heavily on large datasets to generate accurate predictions, data privacy becomes a core ethical issue, requiring strict security protocols and data governance practices.

AI developers and healthcare institutions must adopt privacy-preserving techniques, such as data anonymization, encryption, and differential privacy, to safeguard patient information. However, even anonymized data can sometimes be re-identified when combined with other data sources, presenting a significant challenge to maintaining complete privacy (Sweeney 2013). Regulatory frameworks such as the General Data Protection Regulation (GDPR) in the European Union and the Health Insurance Portability and Accountability Act (HIPAA) in the United States provide guidelines on data privacy, but ensuring compliance remains a complex task. Healthcare providers and AI developers must continuously update data protection practices to stay ahead of evolving threats and ensure patients' right to privacy.

3.2 Algorithmic Bias and Fairness

Algorithmic bias occurs when AI models produce disparate outcomes for different groups, often resulting from biases in the training data. In healthcare, these biases can lead to unequal treatment of specific populations, potentially exacerbating existing health disparities and undermining healthcare equity. For example, studies have shown that some algorithms predict healthcare needs more accurately for white patients than for patients from minority groups due to the under-representation of diverse demographic data in the training datasets (Obermeyer et al. 2019). This bias can result in differential access to care, impacting treatment recommendations, resource allocation, and ultimately, patient outcomes.

Addressing algorithmic bias requires a multifaceted approach, including diversifying training datasets to ensure representation across gender, ethnicity, socioeconomic status, and other demographic factors. Moreover, developers must implement fairness-aware algorithms to detect and correct biases during model training and validation (Mehrabi et al. 2021). Ethical AI development also involves transparency in how models are trained and validated, allowing healthcare providers to understand potential limitations in model outputs and mitigate biases where possible.

3.3 Interpretability and Transparency

Interpretability, or the ability to understand how AI models make predictions, is essential for gaining the trust of healthcare professionals and patients. The 'black-box' nature of many AI models, particularly complex DL algorithms, poses significant ethical concerns in healthcare, where understanding the rationale behind a decision can be as crucial as the decision itself (Doshi-Velez and Kim 2017). For instance, a clinician relying on an AI-driven diagnostic tool may need to understand the basis of a model's recommendation to validate it against clinical judgment and communicate the reasoning to the patient.

Interpretability challenges highlight the ethical need for explainable AI (XAI) models in healthcare, which provide insights into the decision-making process of AI systems. This transparency is crucial in high-stakes situations, where AI recommendations, such as cancer diagnoses or surgical planning, influence treatment decisions. Explainable models also allow clinicians to verify AI outputs, promoting collaboration between AI systems and human expertise (Samek et al. 2017). Research into XAI aims to bridge this gap, developing models that provide explanations of their predictions in a way that clinicians and patients can understand, ultimately supporting more informed and trustworthy healthcare decisions.

3.4 Accountability and Liability

Another ethical consideration in AI adoption is determining accountability and liability when AI-driven systems produce incorrect or harmful outcomes. In healthcare, an AI model's recommendation, if inaccurate, might lead to adverse effects, potentially impacting patient safety and treatment efficacy. Determining who is accountable—the healthcare provider, AI developer, or institution—is complex and presents legal and ethical challenges. This accountability is further complicated by the ability of AI to learn and evolve, leading to recommendations that even its developers may not fully understand (Davenport and Kalakota 2019).

Clear guidelines are necessary to establish liability boundaries for AI-driven healthcare solutions, including frameworks delineating responsibilities among developers, providers, and institutions. Such guidelines should account for the role of AI in decision-making, distinguishing between fully autonomous systems and tools used with human oversight. Furthermore, accountability frameworks should encourage regular audits of AI systems to ensure they operate safely and effectively, as well as mechanisms to address failures and revise algorithms as necessary.

3.5 Patient Consent and Autonomy

AI's reliance on large datasets, often involving patient data, raises ethical questions regarding informed consent. Traditionally, patients consent to data use for specific purposes, but with AI, their data might be reused, combined, or repurposed for different applications than initially intended (Li et al. 2022). Ensuring that patients understand how their data will be used and giving them a choice to opt out are essential requirements for maintaining the autonomy and respect of individuals.

Enhanced consent practices may include transparent communication about the uses of AI in their healthcare, allowing patients to make informed decisions about their participation. Additionally, implementing opt-in policies for using patient data in training and refining AI models is an ethical consideration that respects patient autonomy. This approach requires continuous patient engagement and education, helping patients understand AI's role and the potential impacts on their healthcare.

Ethical considerations in AI for healthcare highlight the need for robust privacy protection, bias mitigation, transparency, accountability, and respect for patient autonomy. Addressing these ethical challenges is crucial to fostering trust in AI systems, ensuring they are not only accurate but also fair, transparent, and respectful of patient rights. This ethical grounding is vital for successful AI adoption in healthcare, as it ensures that AI systems operate in a manner that aligns with the values and needs of both patients and healthcare providers.

4. Data Integration and Quality Management

Data integration and quality management are fundamental to successfully implementing AI in healthcare. AI models rely on comprehensive, high-quality datasets to generate accurate predictions, but healthcare data must be more cohesive, consistent, and diverse. Ensuring data quality and effective integration from multiple sources, such as electronic health records (EHRs), medical imaging, and genomic data, is essential to building robust AI models. This section addresses the primary challenges and solutions associated with integrating diverse data types and maintaining data quality to enable reliable AI applications in healthcare.

4.1 Data Integration Across Sources

Healthcare data is inherently heterogeneous, originating from varied sources such as clinical notes, laboratory results, imaging studies, and wearable devices. Each data type follows different standards, formats, and protocols, making it challenging to combine them into a cohesive dataset suitable for AI analysis (Bates et al. 2018). For instance, EHRs are typically structured in a manner tailored to individual healthcare organizations, which creates discrepancies in data structure, labeling, and categorization. Without standardized formats, integrating data from disparate EHR systems can lead to data loss or misinterpretation, which diminishes the effectiveness of AI models in clinical applications.

To address these challenges, healthcare institutions and data scientists are working towards standardization through initiatives such as Fast Healthcare Interoperability Resources (FHIR), which aims to streamline data exchange between systems (Mandel et al. 2016). FHIR and similar standards support interoperability by defining consistent data formats, facilitating the integration of diverse datasets, and enabling more accurate and timelier AI-driven insights. Additionally, data integration platforms that consolidate data from different sources into unified repositories can streamline data access and preprocessing, enhancing the quality and availability of data for AI models.

4.2 Ensuring Data Quality and Consistency

Data quality is paramount in healthcare, as AI models trained on erroneous or low-quality data can produce unreliable or harmful recommendations. Common data quality issues in healthcare include missing values, redundant records, and inaccurate data entries, which can occur due to human error or variations in data entry practices (Hripcsak and Albers 2013). Addressing these issues requires robust data cleaning and preprocessing pipelines, which identify and resolve inconsistencies to produce high-quality datasets that support accurate and reliable AI predictions.

Data validation and error-checking processes are essential to quality management in healthcare data. For instance, using data imputation techniques to handle missing values and implementing deduplication algorithms to remove redundant records can significantly improve data quality. Additionally, healthcare providers must adopt standardized coding practices, such as the International Classification of Diseases (ICD) codes, to reduce variations and inconsistencies in data labeling (Levine et al. 2018). These practices enhance data consistency, which is vital for building AI models that generalize well across different healthcare settings.

Data quality assurance also extends to maintaining data relevance and timeliness. As healthcare conditions and treatment protocols evolve, outdated data can skew AI predictions, leading to inappropriate recommendations. Implementing mechanisms to update datasets and regularly validate model performance against new data is crucial to maintaining AI applications' long-term accuracy and relevance in healthcare (Miotto et al. 2018, Chen et al. 2020).

4.3 Dealing with Unstructured Data

Much of the data generated in healthcare, such as clinical notes, radiology reports, and pathology results, is unstructured, making it challenging to analyze by using traditional algorithms. This unstructured data often contains valuable insights but requires preprocessing steps, such as natural language processing (NLP), to extract relevant information and convert it into a structured format that AI models can interpret (Shickel et al. 2017). NLP algorithms can process clinical narratives, extracting key information on diagnoses, symptoms, and treatments to enhance the dataset used for model training.

In addition to NLP, imaging data, such as MRIs and CT scans, requires specialized preprocessing, including segmentation, normalization, and artifact removal, to ensure that AI models can interpret the images accurately. By converting unstructured data into a format usable by AI models, healthcare institutions can leverage more comprehensive datasets, enhancing the depth and accuracy of AI-driven insights.

4.4 Data Labeling and Annotation

Data labeling is a critical aspect of quality management in healthcare AI, as labeled data is necessary to train supervised learning models effectively. However, accurate labeling is often challenging in healthcare due to the need for domain expertise and the high cost associated with annotating large datasets. For example, annotating medical images to identify disease markers requires radiologists or trained specialists, which is time-consuming and resource-intensive (Rajpurkar et al. 2017). Inaccurate labeling can also introduce bias and errors into the model, leading to suboptimal performance.

Many researchers are exploring semi-supervised and unsupervised learning techniques and crowdsourcing methods involving clinicians in the annotation process to improve data labeling processes. Additionally, emerging techniques like active learning allow AI models to request labels for only the most informative samples, reducing the overall burden of manual annotation (Settles 2012). These strategies can streamline the labeling process while maintaining accuracy, ultimately contributing to higher-quality datasets for AI models.

4.5 Addressing Data Security in Integrated Systems

Integrating data from multiple sources raises significant security concerns, increasing access points and potentially exposing sensitive healthcare information to breaches. Data security is crucial for maintaining patient trust and complying with regulations like HIPAA and GDPR, which mandate stringent data protection measures. In healthcare AI, ensuring data security involves implementing encryption, access control, and audit mechanisms to protect patient data throughout its lifecycle (Kaissis et al. 2020).

Federated learning is an emerging solution that enables AI models to train on data from multiple sources without transferring raw data between institutions. This approach allows for decentralized learning, preserving patient privacy while allowing institutions to leverage large datasets. By ensuring data security and privacy through federated learning, AI developers can build more comprehensive and reliable models without compromising patient confidentiality (Rieke et al. 2020).

Data integration and quality management are essential for realizing the full potential of AI in healthcare. Addressing data standardization, quality assurance, and security challenges is fundamental to building robust AI models that can improve patient outcomes and support clinical decision-making. As healthcare institutions continue to adopt AI, effective data management practices will play a critical role in ensuring that these technologies provide accurate, trustworthy, and equitable care.

5. Regulatory and Governance Challenges

The integration of AI in healthcare raises complex regulatory and governance issues, given the high stakes involved in patient care, data privacy, and ethical practice. Effective regulation is essential to ensure AI's safe, ethical, and equitable use, but traditional regulatory frameworks often lack the flexibility to accommodate the dynamic nature of AI technologies. This section explores the challenges related to regulatory oversight and governance, highlighting the importance of comprehensive frameworks and guidelines that address the unique needs and risks of AI in healthcare.

5.1 The Need for Comprehensive Regulatory Frameworks

AI's entry into healthcare has outpaced the development of robust regulatory structures, creating a gap between technological advancements and formal governance. Traditional medical device regulations, which focus on physical products and predictable clinical outcomes, struggle to accommodate AI's evolving and data-driven nature. For instance, ML models can continuously learn and update from new data, making it challenging to establish fixed standards for safety and efficacy (McKinney et al. 2020). Without adaptive frameworks, regulatory bodies risk stifling innovation or permitting the deployment of inadequately validated AI systems that could endanger patient safety.

In response to this need, regulatory bodies are beginning to recognize the importance of more adaptive frameworks. The U.S. Food and Drug Administration (FDA) and the European Medicines Agency (EMA), among others, are developing AI-specific guidelines to assess software as a medical device (SaMD) and regulate AI in clinical applications. These frameworks aim to address the lifecycle of AI tools, including their initial approval, post-deployment monitoring, and iterative updates (Benjamens et al. 2020). However, a unified, international approach is still lacking, and developing a consistent regulatory framework that balances innovation with safety remains a pressing challenge.

5.2 Transparency and Accountability

AI models in healthcare often operate as 'black boxes', where complex algorithms make decisions without transparent explanations. This lack of transparency can lead to ethical and legal concerns if an AI-driven decision results in adverse patient outcomes. Determining accountability—whether it lies with the developers, healthcare providers, or institutions—becomes challenging when decision-making is partly or wholly delegated to an opaque algorithm (Davenport and Kalakota 2019). Regulatory frameworks must require AI developers to demonstrate transparency in model functionality, data sources, and potential limitations to build trust and ensure accountability.

Implementing transparency mandates could require AI systems to produce interpretable outputs, such as confidence scores or detailed rationale for clinical recommendations. Additionally, regulations that hold healthcare institutions accountable for the safe deployment of AI systems can ensure these tools undergo rigorous internal testing and monitoring. Without clear guidelines on transparency and accountability, healthcare organizations may face legal liabilities and ethical dilemmas in the event of AI-related errors.

5.3 Ethical Guidelines for Fair and Equitable AI Deployment

One significant regulatory challenge in healthcare AI is ensuring fair and equitable use across diverse patient populations. AI models are susceptible to biases embedded in training datasets, leading to discriminatory outcomes, particularly against minority and underrepresented populations (Obermeyer et al. 2019). Regulatory guidelines must address these concerns by enforcing standards for data diversity, model validation, and bias mitigation to promote equitable healthcare.

Ethical guidelines should encourage using diverse datasets during AI model development to minimize biases that may lead to disparities in treatment recommendations. Regular audits of AI systems can also help identify and address biases that may emerge over time. Regulatory bodies can further incentivize organizations to conduct fairness assessments and adopt bias-reduction strategies, ensuring that AI technologies promote equitable access to high-quality care (Mehrabi et al. 2021). However, achieving standardized and enforceable regulations on fairness and equity remains challenging but necessary.

5.4 Post-Deployment Monitoring and Adaptability

Unlike traditional medical devices, AI models can evolve after deployment, necessitating continuous oversight to ensure ongoing safety and efficacy. Post-deployment monitoring is essential, as AI systems may encounter new data that differs from their training data, potentially affecting performance. Regular updates to AI models can introduce new risks, making it crucial for regulatory bodies to establish guidelines for continuous evaluation and revalidation (Topol 2019). For example, AI models used in diagnostics may need periodic adjustments to account for changing medical knowledge, clinical practices, and demographic shifts in patient populations.

Regulatory frameworks that support post-deployment monitoring involve establishing protocols for tracking model performance, detecting deviations from expected outcomes, and implementing corrective measures as needed. Healthcare institutions must have reporting systems for AI-related adverse events, facilitating transparency and accountability in addressing potential risks. Additionally, adaptive regulatory policies, such as conditional approvals and phased deployment, can enable AI systems to evolve safely without compromising patient safety.

5.5 Standardization and Interoperability Requirements

AI deployment in healthcare often involves integrating models with existing health information systems, such as EHRs, laboratory databases, and imaging platforms. However, the need for standardization and interoperability across these systems poses a significant barrier to seamless AI integration. AI models trained on data from a specific system may perform inconsistently when used with data from different systems, potentially leading to errors in clinical recommendations (Adler-Milstein and Jha 2017). Regulatory bodies must establish interoperability standards that facilitate data exchange and compatibility across platforms.

Standardization efforts, such as adopting Fast Healthcare Interoperability Resources (FHIR) and Health Level Seven (HL7) protocols, play a vital role in achieving seamless data sharing, which is crucial for AI effectiveness. Regulatory frameworks can ensure that AI models receive consistent data inputs by mandating compliance with these interoperability standards, enabling more accurate and reliable predictions across diverse healthcare environments. Ultimately, fostering standardization and interoperability can maximize the impact of AI applications in healthcare by facilitating broader deployment and use.

5.6 Privacy and Data Security Regulations

Given AI's reliance on large datasets, data privacy and security are central regulatory concerns. AI systems require access to extensive amounts of patient data, which increases the risk of privacy breaches and unauthorized data use. Regulations such as GDPR in Europe and HIPAA in the United States mandate strict privacy and security practices to protect patient data (Voigt and Von dem Bussche 2017). However, AI's evolving capabilities present unique challenges that current privacy laws may not fully address, such as re-identification risks associated with anonymized data.

Regulatory bodies are exploring methods such as differential privacy, federated learning, and homomorphic encryption to enhance data security in AI applications, which enable AI models to learn from data while preserving privacy. For example, federated learning allows AI models to train across multiple institutions without transferring patient data, significantly reducing privacy risks (Rieke et al. 2020). Ensuring adherence to privacy and security regulations requires collaboration among regulatory bodies, healthcare providers, and AI developers to implement best practices in data handling and encryption technologies.

Regulatory and governance challenge in healthcare AI is complex and multifaceted. Addressing these challenges requires developing adaptive, transparent, and ethical regulatory frameworks that balance innovation with safety, fairness, and privacy. As AI continues to evolve, ongoing collaboration among healthcare providers, regulatory bodies, and AI developers will be essential to create a regulatory environment that supports AI's responsible and effective use in healthcare.

6. Future Directions for AI in Healthcare

AI in healthcare has already shown transformative potential, but its full realization depends on overcoming current limitations and exploring new avenues for research, application, and regulation. Future directions in AI focus on building more interpretable, equitable, and integrated systems across various healthcare data sources. These advancements aim to create AI solutions that improve diagnostic accuracy and treatment outcomes and ensure safe, ethical, and widespread adoption across diverse healthcare settings. This section examines the most promising future directions in AI for healthcare, including advancements in explainable AI, multimodal data integration, scalability, patient monitoring, and new avenues for clinical research. Some of the key future directions of the AI application in healthcare applications are summarized in Table 2.

Table 2. Future directions for AI in healthcare.

Direction	Description	Potential Impact
Explainable AI (XAI)	Developing models that are interpretable and transparent.	Enhances clinician trust and shared decision-making.
Multimodal Data Integration	Combining diverse data types (e.g., imaging, genomics).	Enables more personalized and precise treatment plans.
Ethical AI Development	Focusing on fairness-aware algorithms and diverse datasets.	Reduces healthcare disparities and builds public trust.
Scalable AI Models	Creating lightweight models suitable for diverse healthcare settings.	Expands AI use in resource-constrained environments.

6.1 Advancements in Explainable and Trustworthy AI

One of the major barriers to AI adoption in healthcare is the 'black box' nature of many AI models, which makes it difficult for clinicians to understand the decision-making process behind AI-generated recommendations. The development of explainable AI (XAI) is crucial in addressing this challenge. Explainable AI techniques, such as feature importance rankings and visual explanations, allow clinicians to understand which factors contribute most to the AI's predictions (Doshi-Velez and Kim 2017). Such transparency builds clinician trust and facilitates more informed, shared decision-making between patients and healthcare providers.

Trustworthy AI also involves developing systems that can provide confidence estimates for their predictions, enabling clinicians to assess better when an AI model is reliable or when human oversight is essential. Future research will continue enhancing interpretability through model distillation, translating complex model outputs into simpler, more understandable representations (Samek et al. 2017). These advancements aim to bridge the gap between AI and human clinicians, creating tools that support rather than replace clinical expertise.

6.2 Integration of Multimodal Data

A significant future direction for AI in healthcare is integrating multimodal data to create more comprehensive and accurate models. Healthcare data is not only diverse, encompassing imaging, genomics, clinical notes, and more, but also often siloed across different systems. Multimodal AI combines multiple data types and has shown promise in yielding richer insights by capturing complex relationships among various data sources. For example, combining imaging data with genetic profiles and clinical records could enable more precise diagnoses and tailored treatment plans for complex diseases like cancer and cardiovascular disorders (Miotto et al. 2018, Chen et al. 2020).

Integrating multimodal data requires advancements in data harmonization techniques and standardized formats and interoperability protocols, such as FHIR and HL7, which facilitate seamless data exchange. Future AI systems will likely leverage more advanced techniques like transfer learning to adapt multimodal models across varied healthcare environments and improve performance on limited datasets. This integration will play a crucial role in driving personalized medicine, where treatment can be adapted to individual patient profiles based on a holistic view of their health data.

6.3 Scalability and Deployment in Diverse Settings

Scalability is a pressing issue in healthcare AI, especially given the need to adapt AI tools to different clinical environments with varying resources and patient demographics. Current AI models often require extensive computational power, limiting their deployment in resource-constrained settings. Future AI solutions will focus on creating lightweight, efficient models that maintain accuracy while reducing computational demands, making them suitable for deployment in diverse settings, including rural and underserved areas (Rajpurkar et al. 2017).

In addition to developing resource-efficient models, achieving scalability requires standardized protocols for deploying, monitoring, and updating AI systems. Real-time monitoring frameworks can help track AI model performance and detect potential drifts in accuracy when applied to new data, ensuring that models remain reliable over time. By creating scalable AI systems, healthcare organizations can expand the reach of AI to improve access to quality care for populations that may lack immediate access to advanced healthcare facilities.

6.4 Enhanced Clinical Trial Design and Patient Monitoring

AI offers promising potential in reshaping clinical trials, making them more efficient, targeted, and patient-centered. In traditional clinical trials, patient recruitment, monitoring, and data analysis are time-consuming and costly. Future AI applications could enhance clinical trial design by using predictive analytics to identify eligible participants more quickly, optimizing trial protocols based on real-time data, and improving patient adherence and engagement (Esteva et al. 2019).

AI-driven patient monitoring through wearable devices and remote health technologies enables the continuous collection of real-world health data, which can be invaluable in detecting early signs of adverse events or treatment responses. Real-time patient data allows researchers to adapt clinical trials dynamically, reducing participant burden and enhancing the quality of collected data. This application not only streamlines clinical trials but also supports the development of AI-driven personalized treatment recommendations based on continuous patient monitoring (Topo 2019).

6.5 Development of Ethical and Fair AI Systems

The future of AI in healthcare must prioritize fairness and equity to prevent the exacerbation of healthcare disparities. Future AI systems must incorporate fairness-aware algorithms to detect and mitigate biases during model training and application. This will likely include developing bias auditing tools and standards for measuring fairness in clinical AI systems, ensuring that AI models do not produce skewed outcomes that disproportionately affect underrepresented populations (Mehrabi et al. 2021).

Ensuring fairness also involves ongoing efforts to diversify training datasets, incorporating data from varied demographic and socioeconomic groups. Researchers and healthcare institutions are increasingly aware of the need for ethical AI, leading to collaboration across disciplines to develop standards and guidelines that promote inclusivity and fairness. As these standards evolve, they will help foster AI systems that provide equitable and high-quality care for all patients, regardless of background or identity.

6.6 Regulatory Innovation and Adaptive Policy Frameworks

AI's rapid evolution necessitates regulatory frameworks that can adapt to emerging technologies and evolving clinical needs. Current regulatory processes are often rigid and slow, limiting their ability to accommodate AI's iterative nature. In the future, regulatory bodies must adopt more flexible, adaptive policies allowing AI's continuous learning and updating capabilities. This might include conditional approvals, expedited regulatory pathways, and post-deployment monitoring systems that enable AI models to be refined and improved even after deployment (Benjamens et al. 2020).

Innovative frameworks such as 'sandbox' environments, where AI applications can be tested in controlled, real-world settings, could support the safe exploration of new AI technologies before broader implementation. Collaborative approaches involving healthcare providers, regulatory agencies, AI developers, and ethicists will be critical in crafting policies that enable innovation while safeguarding patient safety and privacy. Adaptive regulation will ultimately help create a balanced environment where AI can thrive responsibly within healthcare.

The future of AI in healthcare lies in creating systems that are explainable, scalable, equitable, and aligned with ethical standards. Integrating multimodal data, advancements in explainability, enhanced clinical trial design, and the development of adaptive regulatory frameworks will shape the trajectory of AI in healthcare. With these advancements, AI can evolve into a transformative force that not only supports but enhances the quality and accessibility of healthcare worldwide.

7. Conclusion

Artificial Intelligence (AI) in healthcare holds transformative potential, promising to improve diagnostic accuracy, enable personalized treatments, and enhance overall patient outcomes. However, realizing AI's full potential in healthcare requires addressing several challenges related to data integration, model reliability, ethical standards, and regulatory frameworks. This chapter has explored the complex landscape of AI in healthcare, emphasizing the importance of developing adaptive, transparent, and ethical AI systems that align with the healthcare sector's unique demands.

The adoption of AI in healthcare encounters significant obstacles, including the technical challenges of integrating diverse data sources, ensuring data quality, and building robust models that generalize across varied patient populations. Data privacy, algorithmic fairness, and model interpretability remain critical ethical considerations, as the healthcare sector must balance technological innovation with the rights and welfare of patients. Addressing these issues is crucial for ensuring that AI systems provide accurate and equitable care and fostering trust among healthcare providers and patients.

Regulatory and governance challenges further complicate AI integration in healthcare. Existing frameworks often lack the flexibility to keep pace with the rapid evolution of AI technologies, underscoring the need for adaptive policies that can accommodate continuous learning, post-deployment monitoring, and iterative improvements. Establishing standards for data interoperability, accountability, and fairness is essential for creating a regulatory environment that supports safe and effective AI applications. Collaborations among regulatory bodies, healthcare providers, and AI developers will be vital for developing policies that protect patient safety while encouraging innovation.

Future directions in healthcare AI focus on creating scalable, explainable, and equitable systems that can be deployed across diverse healthcare settings. Advancements in explainable AI, multimodal data integration, and enhanced clinical trial designs are expected to drive AI's growth, fostering a new era of personalized medicine. Additionally, as regulatory bodies embrace adaptive frameworks, AI will have more flexibility to evolve responsibly, meeting the needs of a wide range of healthcare applications.

The successful integration of AI in healthcare will depend on a balanced approach that addresses technical, ethical, and regulatory challenges. By fostering collaboration among healthcare professionals, data scientists, policymakers, and ethicists, the healthcare sector can create AI systems that are safe, reliable, and fair. As AI technology continues to advance, it has the potential to reshape healthcare in unprecedented ways, ultimately contributing to a more equitable, efficient, and patient-centered healthcare system.

References

Adler-Milstein, Julia and Ashish K. Jha. (2017). HITECH Act drove large gains in hospital electronic health record adoption. Health Affairs 36(8): 1416–1422.

Bates, David W., Suchi Saria, Lucila Ohno-Machado, Anand Shah and Gabriel Escobar. (2014). Big data in health care: using analytics to identify and manage high-risk and high-cost patients. Health Affairs 33(7): 1123–1131.

Benjamens, Stan, Pranavsingh Dhunnoo and Bertalan Meskó. (2020). The state of artificial intelligence-based FDA-approved medical devices and algorithms: an online database. NPJ Digital Medicine 3(1): 118.

Birkhead, Guthrie S., Michael Klompas and Nirav R. Shah. (2015). Uses of electronic health records for public health surveillance to advance public health. Annual Review of Public Health 36(1): 345–359.

Chen, Yen-Wei and Lakhmi C. Jain. (2020). Deep Learning in Healthcare. Paradigms and Applications; Springer: Berlin/Heidelberg, Germany.

Davenport, Thomas and Ravi Kalakota. (2019). The potential for artificial intelligence in healthcare. Future Healthcare Journal 6(2): 94–98.

Doshi-Velez, Finale and Been Kim. (2017). Towards a rigorous science of interpretable machine learning. arXiv preprint arXiv:1702.08608.

Esteva, Andre, Alexandre Robicquet, Bharath Ramsundar, Volodymyr Kuleshov, Mark DePristo, Katherine Chou et al. (2019). A guide to deep learning in healthcare. Nature Medicine 25(1): 24–29.

Hripcsak, George and David J. Albers. (2013). Next-generation phenotyping of electronic health records. Journal of the American Medical Informatics Association 20(1): 117–121.

Jiang, Fei, Yong Jiang, Hui Zhi, Yi Dong, Hao Li, Sufeng Ma et al. (2017). Artificial intelligence in healthcare: past, present and future. Stroke and Vascular Neurology 2(4).

Johnson, Alistair E. W., Tom J. Pollard, Lu Shen, Li-wei H. Lehman, Mengling Feng, Mohammad Ghassemi et al. (2016). MIMIC-III, a freely accessible critical care database. Scientific Data 3(1): 1–9.

Kaissis, Georgios A., Marcus R. Makowski, Daniel Rückert and Rickmer F. Braren. (2020). Secure, privacy-preserving and federated machine learning in medical imaging. Nature Machine Intelligence 2(6): 305–311.

Kohane, Isaac S. (2015). Ten things we have to do to achieve precision medicine. Science 349(6243): 37–38.

Levine, David M., Kei Ouchi, Bonnie Blanchfield, Keren Diamond, Adam Licurse, Charles T. Pu et al. (2018). Hospital-level care at home for acutely ill adults: a pilot randomized controlled trial. Journal of General Internal Medicine 33(5): 729–736.

Li, Fan, Nick Ruijs and Yuan Lu. (2022). Ethics & AI: A systematic review on ethical concerns and related strategies for designing with AI in healthcare. AI 4(1): 28–53.

Mandel, Joshua C., David A. Kreda, Kenneth D. Mandl, Isaac S. Kohane and Rachel B. Ramoni. (2016). SMART on FHIR: a standards-based, interoperable apps platform for electronic health records. Journal of the American Medical Informatics Association 23(5): 899–908.

McKinney, Scott Mayer, Marcin Sieniek, Varun Godbole, Jonathan Godwin, Natasha Antropova, Hutan Ashrafian et al. (2020). International evaluation of an AI system for breast cancer screening. Nature 577(7788): 89–94.

Mehrabi, Ninareh, Fred Morstatter, Nripsuta Saxena, Kristina Lerman and Aram Galstyan. (2021). A survey on bias and fairness in machine learning. ACM Computing Surveys (CSUR) 54(6): 1–35.

Miotto, Riccardo, Fei Wang, Shuang Wang, Xiaoqian Jiang and Joel T. Dudley. (2018). Deep learning for healthcare: review, opportunities and challenges. Briefings in Bioinformatics 19(6): 1236–1246.

Obermeyer, Ziad, Brian Powers, Christine Vogeli and Sendhil Mullainathan. (2019). Dissecting racial bias in an algorithm used to manage the health of populations. Science 366(6464): 447–453.

Rajpurkar, P. (2017). CheXNet: Radiologist-Level Pneumonia Detection on Chest X-Rays with Deep Learning. ArXiv abs/1711 5225.

Ravi, Daniele, Charence Wong, Benny Lo and Guang-Zhong Yang. (2016). A deep learning approach to on-node sensor data analytics for mobile or wearable devices. IEEE Journal of Biomedical and Health Informatics 21(1): 56–64.

Rieke, Nicola, Jonny Hancox, Wenqi Li, Fausto Milletari, Holger R. Roth, Shadi Albarqouni et al. (2020). The future of digital health with federated learning. NPJ Digital Medicine 3(1): 1–7.

Samek, W. (2017). Explainable artificial intelligence: Understanding, visualizing and interpreting deep learning models. arXiv preprint arXiv:1708.08296.

Settles, B. (2012). Active Learning (1st ed.). Springer Cham. https://doi.org/10.1007/978-3-031-01560-1.

Shickel, Benjamin, Patrick James Tighe, Azra Bihorac and Parisa Rashidi. (2017). Deep EHR: a survey of recent advances in deep learning techniques for electronic health record (EHR) analysis. IEEE Journal of Biomedical and Health Informatics 22(5): 1589–1604.

Shortliffe, Edward H. and Martin J. Sepúlveda. (2018). Clinical decision support in the era of artificial intelligence. Jama 320(21): 2199–2200.

Sweeney, Latanya. (2013). Discrimination in online ad delivery. Communications of the ACM 56(5): 44–54.

Topol, Eric J. (2019). High-performance medicine: the convergence of human and artificial intelligence. Nature Medicine 25(1): 44–56.

Varoquaux, Gaël. (2018). Cross-validation failure: Small sample sizes lead to large error bars. Neuroimage 180: 68–77.

Voigt, Paul and Axel Von dem Bussche. (2017). The EU general data protection regulation (GDPR). A Practical Guide, 1st Ed., Cham: Springer International Publishing 10(3152676): 10–5555.

Yu, Kun-Hsing, Andrew L. Beam and Isaac S. Kohane. (2018). Artificial intelligence in healthcare. Nature Biomedical Engineering 2(10): 719–731.

Index

P

S

T

About the Authors

S. Vijayalakshmi

Dr. S. Vijayalakshmi is an Associate Professor in the School of Sciences, at Christ University, Pune Lavasa Campus, where she serves in multiple academic portfolios. Her research expertise encompasses Image Processing and Internet of Things (IoT), areas in which she has made significant contributions through publications in numerous international and national conferences as well as reputed journals. Dr. S. Vijayalakshmi has authored several book chapters and currently has multiple books in development. As an active member of academic bodies such as IFERP, IAENG and IEEE she remains engaged with the broader research community while dedicating considerable effort to guiding students in both research and academic pursuits. Her commitment to advancing knowledge in data science and emerging technologies, combined with her passion for student mentorship, makes her a valuable contributor to both academia and the field of technological innovation.

Alwin Joseph

Mr. Alwin Joseph is an Assistant Professor at CHRIST University, Pune, Lavasa Campus. He holds a postgraduate degree in Computer Applications from CHRIST, Bangalore, and is currently pursuing his Ph.D. in Computer Science. His research interests encompass Image Processing, Machine Learning, and Natural Language Processing. Mr. Alwin complements his academic pursuits with practical experience from live projects and a full-stack application developer internship. He is proficient in various programming and scripting languages, including C, C++, JAVA, Python, and JavaScript, and is passionate about leveraging these skills in his teaching and research.

Naived George Eapen

Naived George Eapen is a Post Graduate in Computer Science and a Google Certified Educator with hands-on experience in AI and ML, working as an Assistant Professor at CHRIST University, Pune, Lavasa Campus. He has a proven track record through internships as both a Full Stack Web Developer and an AI Engineer, contributing to live projects in the field. His primary interests lie in IoT, Computer Vision, Machine Learning, and API Development. Mr. Naived is proficient in a wide range of programming and scripting languages, including C, JAVA, Python, and JavaScript, and is passionate about applying these skills to developing innovative solutions.

Balamurugan Balusamy

Dr. Balamurugan Balusamy is the Associate Dean at Shiv Nadar Institution of Eminence, Delhi-NCR, and an Adjunct Professor at Taylor's University, Malaysia. With a Ph.D. from a premier Indian institution and 12 years of experience at VIT University, his expertise lies in Blockchain and Data Sciences. Formerly a Professor and Director at Galgotias University, he is a prolific researcher with over 200 high-impact papers and has authored or edited more than 80 books. A passionate educator and consultant for Industrial IoT, he has delivered over 195 talks globally.

Jagjit Singh Dhatterwal

Dr. Jagjit Singh Dhatterwal is presently working as an Associate Professor with the Department of School of Computer Science & Artificial Intelligence, SR University, Warangal, Telangana, India. He has supervised many UG and PG projects of engineering students. He is also a Member of Computer Science Teacher Association (CSTA), New York, USA, International Association of Engineers (IAENG), Hong Kong, IACSIT (International Association of Computer Science and Information Technology), USA, Professional Member of the Association of Computing Machinery, USA, IEEE, and a Life Member of the Computer Society of India. He has published more than 15 books and 120+ book chapters at National/International level. He also has more than 20 publications in International/National Journal and conferences. His areas of interest include Artificial Intelligence, BCI, Cyborg and Multi-Agents Technology. He has number of publications as Books/Book chapters/Journal papers and conference papers.

Kuldeep Singh Kaswan

Professor (Dr.) Kuldeep Singh Kaswan is presently working in the School of Computer Science & Engineering, Galgotias University, Uttar Pradesh, India. His contributions focus on BCI, Cyborgs, and Data Science. His academic degrees and thirteen years of experience working with global Universities like Amity University, Noida, India; Gautam Buddha University, Greater Noida; and PDM University, Bahadurgarh, India, have made him more receptive and prominent in his domain. He received a doctorate in Computer Science from Banasthali Vidyapith, Rajasthan. He received a Doctor of Engineering (D. Engg.) from Dana Brain Health Institute, Iran. He has supervised many UG and PG projects by engineering students. He has supervised four Ph.D. graduates. He is also a senior member of IEEE, the Computer Science Teacher Association (CSTA), New York, USA, the International Association of Engineers (IAENG), Hong Kong, and IACSIT, USA. He has published more than 20 books and 150+ book chapters at National/International level. He also has more than 40 publications in International/National Journal and conferences. He is an editor/author and review editor of Journals and Books with IEEE, Wiley, Springer, IGI, River, etc.